This is a comprehensive account of RR Lyrae stars, and traces the story from their initial discovery a century ago, through to their present status. RR Lyrae stars are increasingly important as astronomical tools, such as their use as standard candles for determining stellar distances. This book reviews our current understanding of RR Lyrae variable stars. It is a unique explanation of the multiple applications of these variable stars for a range of astrophysical problems. The author describes the use of RR Lyrae stars as probes of old stellar populations, both in the Milky Way and other galaxies, and as an outstanding testing ground for stellar evolution and pulsation theories.

The author stresses the significance of variable stars for our ultimate understanding of the history and scale of the Milky Way and nearer extragalactic systems. For advanced students and researchers of astronomy, this is a definitive account of the modern theories surrounding RR Lyrae variable stars.

Cambridge astrophysics series

Series editors

Andrew King, Douglas Lin, Stephen Maran, Jim Pringle, and Martin Ward

RR LYRAE STARS

HORACE A. SMITH

Department of Physics and Astronomy, Michigan State University

CAMBRIDGE
UNIVERSITY PRESS

PUBLISHED BY THE PRESS SYNDICATE OF THE UNIVERSITY OF CAMBRIDGE
The Pitt Building, Trumpington Street, Cambridge, United Kingdom

CAMBRIDGE UNIVERSITY PRESS
The Edinburgh Building, Cambridge CB2 2RU, UK
40 West 20th Street, New York NY 10011–4211, USA
477 Williamstown Road, Port Melbourne, VIC 3207, Australia
Ruiz de Alarcón 13, 28014 Madrid, Spain
Dock House, The Waterfront, Cape Town 8001, South Africa

http://www.cambridge.org

First published 1995
First paperback edition 2003

A catalogue record for this book is available from the British Library

Library of Congress cataloguing in publication data

Smith, Horace A.
RR Lyrae stars / Horace A. Smith.
  p.   cm. — (Cambridge astrophysics series ; 27)
ISBN 0 521 32180 8 hardback
1. RR Lyrae stars.   I. Title.   II. Series.
QB843.R72S65   1995
523.8\4425—dc20   94-31146 CIP

ISBN 0 521 32180 8 hardback
ISBN 0 521 54817 9 paperback

Contents

Preface

This review of the RR Lyrae stars is written from the perspective of an observational astronomer. Without, I hope, neglecting the vital contributions of stellar evolution theory and stellar pulsation theory to the present appreciation of these variables, I have nonetheless tried to keep the emphasis of each chapter upon the observations which underlie our understanding of them. I offer my apologies to those theorists who may feel slighted by this approach.

Still, even observational astronomers may be disgruntled over at least one aspect of the text. Whatever knowledge we have of the RR Lyrae stars is based upon literally millions of observations of thousands of these variables in globular clusters, in the galactic field, and in systems beyond the bounds of the Milky Way. Many different observers have contributed in one way or another to the accumulation of these observations. There is no way to credit all students of these stars, whether they be observers or theorists, nor is there space to reference every relevant paper. Often reference has been made to just one or two salient papers in a particular field, or to papers summarizing a large corpus of work. I hope that any dissatisfaction thereby incurred is at least partially assuaged by this general acknowledgement of the efforts of numerous researchers, unnamed in this text, without whose endeavors our knowledge of the RR Lyrae stars would be much the poorer.

I was assisted by many people in the preparation of this book to whom many thanks are owed. I thank Timothy Beers, Suzanne Hawley, Nancy Silbermann, and Amelia Wehlau for reading and commenting upon draft chapters of this work. Dorrit Hoffleit made available her expertise on the history of variable star astronomy and communicated valuable information about the discovery of the first RR Lyrae variable. The staffs of the libraries of the Physics and Astronomy Department at Michigan State University and of the Dominion Astrophysical Observatory were of great help. I thank, in particular, Diane Clark and Judy Matthews for their assistance in tracking down obscure references. I thank the staff of the Dominion Astrophysical Observatory and its director, Jim Hesser, for their hospitality while I was on sabbatical leave from Michigan State University. Summer student Brian Kern assisted in the tabulation of data on field and cluster RR Lyrae stars and Tom McWilliams assisted in the sorting of references. Debbie Benedict provided valuable secretarial aid. I thank Simon Mitton for his patience during the preparation of this work. The cooperation of those who gave permission for the reproduction of previously published figures is greatly appreciated. Finally, I thank my mother and members of my family for their encouragement.

1

Introduction

The study of RR Lyrae stars is now a century old. In the closing decade of the nineteenth century, as astronomers subjected the globular clusters to increasingly close scrutiny, they discovered the first of the short-period variable stars today known as RR Lyrae stars. In the course of the ensuing hundred years, the ranks of the RR Lyraes have swelled so that in representatives they outnumber known members of any other well-defined class of variable. Directly or indirectly, investigations of RR Lyrae stars have contributed to almost every branch of modern astronomy:

- RR Lyrae stars have been tracers of the chemical and dynamical properties of old stellar populations within our own and nearby galaxies;
- RR Lyrae stars have served as standard candles, indicating the distances to globular clusters, to the center of the Galaxy, and to neighboring Local Group systems;
- RR Lyrae stars have served as test objects for theories of the evolution of low-mass stars and for theories of stellar pulsation.

In recent years, these and other applications have helped to make the study of RR Lyrae stars a particularly active field and one which seems likely to continue so for the forseeable future.

This introductory chapter has two purposes: first, to provide a brief historical review of the recognition of RR Lyrae stars as a distinct class of variable star, and second, to summarize the salient characteristics of this type of variable. While those already somewhat acquainted with the RR Lyrae stars will find much in this chapter familiar, those with little prior knowledge of these variables will find in it the background needed to appreciate the more specialized topics of subsequent chapters.

1.1   Recognition of RR Lyrae stars as a distinct class of variable star

The latest edition of the *General Catalogue of Variable Stars* (Kholopov et al. 1985) defines RR Lyrae stars as 'radially pulsating giant A–F stars with periods in the range 0.2–1.2 day and light amplitudes from 0.2 mag to 2 mag [in] V'. Underlying this brief definition are decades of work, commencing, like so much else in the study of variable stars, with the application of photography to the study of the heavens.

The discovery of the first RR Lyrae stars is intimately connected with the realization that some of the stars in globular star clusters are variable stars. The first globular cluster variable to be discovered was a nova which erupted in the cluster M80 in the year 1860. Almost three decades later, in 1889, E. C. Pickering reported the discovery

1

of a second globular cluster variable, a bright variable star near the center of M3. During the following years a few more of the brighter globular cluster variables were found. However, the main story of globular cluster variables, and also of RR Lyrae stars, began in 1893, when Solon I. Bailey initiated a program of globular cluster photography at the Harvard College Observatory station in Arequipa, Peru. In August of that year Williamina Fleming found on Arequipa plates a variable star in the globular cluster ω Centauri. A few days later Pickering found a second variable in the same cluster, and at about the same time, Bailey himself found three variable stars in 47 Tucanae. The discovery of these first Arequipa variables stimulated further searches for variable stars within globular clusters.

In February 1895, Pickering detected six more variables in the globular cluster ω Centauri. That proved to be the start of a flood of discoveries. Between 1895 and 1898 Bailey (1913) searched photographs of 23 globular clusters for variables, discovering more than 500. By the end of this survey, Bailey had discovered almost as many variable stars in globular clusters as had been discovered throughout the remainder of the sky. Variable stars were, however, not equally plentiful in every globular cluster. Bailey noted that, while some globular clusters such as ω Cen, M3, and M5 contained large numbers of variables, others contained few if any variable stars.

Bailey set himself the task of determining periods and lightcurves for the globular cluster variables and, as his work proceeded, it became clear that most of these variables were of a particular type. These variables had short periods of under a day and photographic amplitudes of about 1 magnitude. Moreover, the mean apparent magnitudes of the short-period variables within any given globular cluster were all about the same. These 'cluster-type variables', as they came to be known, were what we should today call RR Lyrae stars.

What was actually the first RR Lyrae star identified? E. C. Pickering's 1889 variable in M3 was probably a Cepheid rather than an RR Lyrae star (Hoffleit 1993). The first RR Lyrae known to occur in a globular cluster may have been found not by Harvard researchers, but by D. E. Packer in 1890. Packer (1890) visually discovered two variable stars in M5, though he did not determine their periods. One of these is now known to be a 26 day period type II Cepheid, but the identity of the second is unclear. E. E. Barnard thought that it, too, was a star later revealed as a Cepheid, but Bailey suspected that Packer had observed a blend of three stars, one of which is an RR Lyrae variable (Bailey 1917). The first RR Lyrae star to be identified outside a cluster may be U Leporis, discovered by J. C. Kapteyn (1890; Hoffleit 1993) and later recognized as a 0.58 day period variable. The paper reporting Kapteyn's discovery was submitted for publication eight days before Packer's first submission. Thus, even were Packer's second variable an RR Lyrae star, technical credit for discovery of the first RR Lyrae variable must go to Kapteyn. The question of who found the first RR Lyrae is, however, mainly a point of curiosity, since whichever was the first RR Lyrae to be found, in consideration of the magnitude of his endeavors, real credit for discovering the class of RR Lyrae stars must go to Bailey. The significance of U Leporis would not be apparent until after Bailey's discoveries of RR Lyrae variables within the globular clusters.

In his paper discussing the variable stars in ω Cen, Bailey (1902) divided the cluster-type variables into three subclasses: Bailey types a, b, and c (figure 1.1). Lightcurve shape was his principal classification criterion. Because these subclasses are still in use

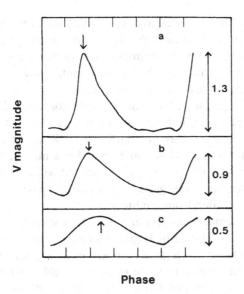

Figure 1.1 The differing lightcurve shapes of RR Lyrae stars of Bailey types a, b, and c.

(though usually simplified to just two types, ab and c, – hereinafter, RRab and RRc, see §1.2.4) it is worth quoting Bailey's original description:

> Subclass a . . . Increase of light very rapid. Decrease rapid, but much less rapid than the increase. Light nearly constant at minimum for about one half of the full period, but perhaps during this time the light changes slowly. In this cluster the range is generally a little more than a magnitude, and the period from twelve to fifteen hours.
>
> Subclass b . . . Increase of light moderately rapid. Decrease is relatively slow and continues with lessening rapidity till about the beginning of increase, except that in some cases there is a tendency to a 'stillstand.' In this cluster, the range is generally a little less than a magnitude, and the period from fifteen to twenty hours . . . This subclass is similar to a, of which it may be regarded as a modification.
>
> Subclass c . . . Light appears to be always changing, and with moderate rapidity. Increase of light generally somewhat more rapid than the decrease, but in a few cases it appears to be of only equal, or of less rapidity. In this cluster the range is generally somewhat more than half a magnitude, and the period from eight to ten hours . . .

Although Bailey was discovering hundreds of RR Lyrae stars in globular clusters, at first few examples of this class were known outside of clusters, in the general field of the Galaxy. The discovery of U Leporis was followed by that of another field RR Lyrae, S Arae, in 1898. Most importantly, W. Fleming, sometime prior to July 1899, discovered a seventh magnitude variable in Lyra which had a period of 0.56 day (Pickering 1901). Pickering noted that in its lightcurve and period this star was indistinguishable from the 'cluster-type' variables. It was given the designation RR Lyrae and it remains the brightest known member of the class. Slowly at first, and then with increasing frequency, more and more of these field 'cluster-type' variables were found. They were initially very important in part because they were nearer and brighter than the variables in globular clusters and so could be studied by techniques such as spectroscopy, which could not then be applied to their faint cluster counterparts. Gradually, the number of field cluster-type stars increased until more of these variables were known in the general field than within globular clusters. As more and more examples were identified outside of globular clusters, the name 'cluster-type' variable no longer seemed appropriate, and the competing term of RR Lyrae star began to be used. Occasionally, other nomenclature is encountered in the early literature, for example, 'antalgol stars', since unlike eclipsing binaries of the Algol-type, many RR Lyraes spend most of their time near minimum rather than maximum light. It was not until the 1948 meeting of Commission 27 of the International Astronomical Union in Zurich that a motion was made by A. H. Joy that 'cluster-type or RR Lyrae-type variables should be called RR Lyrae stars, whether they occur in clusters or in our galaxy'. This motion passed and the name cluster-type variables thereafter faded from the literature.

By about 1915, sufficient data had accumulated concerning the field and cluster RR Lyrae stars to raise the issue of whether they should be regarded as a separate class of variable or included with the Cepheids. Though their periods were shorter, the lightcurve shapes of the RR Lyraes were similar to those of the Cepheid variables. Moreover, spectroscopic studies, especially those of Kiess (1912), indicated that the radial velocity of an RR Lyrae star varied during the course of its light cycle in the same general way as did the radial velocity of a Cepheid. The correct deduction was made, that the basic mechanism of variability was the same for RR Lyrae stars and Cepheids, though at the time of Kiess's study most astronomers still believed that Cepheids were spectroscopic binary stars and that their variability was somehow related to their binary nature. Shapley (1914), in a paper advancing the idea that pulsation was really responsible for the periodic light and radial velocity variations of the Cepheids, concluded that pulsation must also be responsible for the variability of the RR Lyraes. The similarities between RR Lyrae stars and Cepheids encouraged Shapley (1918) to incorporate RR Lyrae stars in his calibration of the Cepheid period–luminosity relation. As tools for measuring the distances to globular clusters, the RR Lyrae stars played a vital role in Shapley's epochal determination of the distance to the center of the Galaxy.

Still, there were reasons for maintaining RR Lyraes as a separate class of variable, rather than subsuming them into the class of Cepheids. Their short periods and their abundance in some globular clusters were, of course, marks of distinction. Hertzsprung (1909; 1913) first pointed out another. Among the field variables, most Cepheids – those which we should now call type I or classical Cepheids – were concentrated to the plane of the Milky Way. The RR Lyraes, on the other hand, were found at all galactic latitudes. In addition, it was found that, unlike most Cepheids, many field RR Lyraes had high radial velocities, a result later confirmed by Joy (1938) from more extensive Mount Wilson observations. These properties we now recognize as indicating that, whereas the classical Cepheids belong to Population I, the RR Lyraes include halo stars of Population II. The high RR Lyrae radial velocities vitiated the argument of Kapteyn and van Rhijn (1922) that the occurrence of field RR Lyraes at all galactic latitudes and their relatively large proper motions implied that the field RR Lyrae stars were faint, relatively nearby, dwarf stars.

Though Eddington (1926) included RR Lyrae in a table of important Cepheid variables in his influential book *The Internal Constitution of the Stars,* others, such as Russell (1927), drew the distinction between Cepheids and RR Lyraes more sharply.

Thus, despite similarities with pulsating Cepheid variables of period longer than a day, the RR Lyrae stars from the first decades of this century were usually looked upon as a distinct class of variable, though one having considerable affinity to the Cepheids. More recently, the realization that all RR Lyrae stars are low-mass horizontal branch stars in the core helium burning stage of evolution has provided an additional argument for distinguishing them from the higher mass classical Cepheids.

There have, however, been adjustments from time to time in the types of variable star included in the class of RR Lyrae stars. A particular confusion arose with the short period pulsating variable stars which are now usually called δ Scuti stars (if Pop. I) or SX Phoenicis stars (if Pop. II), but which have sometimes also been termed dwarf Cepheids, RRs variables, AI Velorum stars, or ultrashort period Cepheids. Reviews of these types of variable star have recently been given by Nemec and Mateo (1990) and Breger (1990). These variables occur within the instability strip below the level of the horizontal branch, near the location of the main sequence. Their periods are short, less than 0.2 days, and their amplitudes in V are usually small (but not invariably so). At first, it seemed natural to include these short-period pulsating stars among the RR Lyraes. However, after it was gradually realized that SX Phe and δ Scuti variables differ both in evolutionary state and absolute magnitude from the classical 'cluster-type' variable, it seemed desirable to clearly separate the SX Phe and δ Scuti stars from the group of RR Lyrae variables. The SX Phe and δ Scuti variables will therefore not be discussed further here, and those interested in them are directed to the reviews mentioned above. The location of the RR Lyrae stars in the HR diagram relative to some other well defined classes of variable stars is shown in figure 1.2.

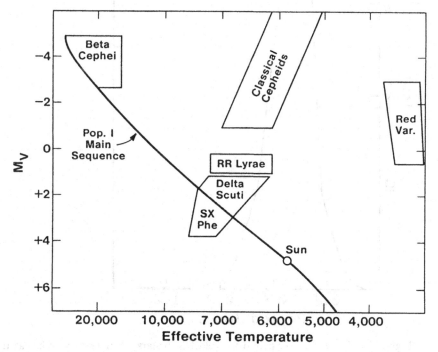

Figure 1.2 The approximate locations in the HR diagram of the RR Lyrae stars and several other well defined classes of variable star.

1.2   Survey of current knowledge

1.2.1   *Numbers of RR Lyrae stars*

It has been evident since Bailey's pioneering studies that RR Lyrae stars are a relatively common class of variable star. By the late 1980s some 1900 RR Lyrae stars had been discovered in 77 globular clusters of our Galaxy (Suntzeff, Kinman, and Kraft 1991). More than 90 percent of known globular cluster variables are RR Lyrae stars. RR Lyraes are also well represented among the field variables of the Galaxy. These field RR Lyraes are found in the halo of the Galaxy, in the thick disk, and in the galactic bulge. The *Bibliographic Catalogue of RR Lyrae Stars* by Heck (1988) includes data on 6367 RR Lyraes in the general field, most of which are also listed in the fourth *General Catalogue of Variable Stars* (Kholopov et al. 1985). More than a fifth of the 28 457 variable stars listed in this catalog are RR Lyrae stars. More than seven RR Lyrae stars are known for each classical Cepheid or W Virginis star which has been discovered in the Galaxy. Reasonably good lightcurves and periods are available for about 4280 field RR Lyrae stars. Of these, about 91 percent are of type RRab and only 9 percent of type

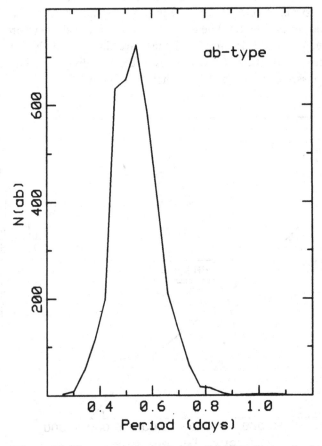

Figure 1.3 Histogram over period for field RRab stars in the Galaxy. Whereas there may be a few RRab stars of greater or lesser period, most have periods between 0.3 and 0.9 days. From Novikova (1988).

RRc. This, however, may not reflect the actual ratio of RRab to RRc variables in the galactic field: the smaller amplitude RRc stars are less likely to be discovered in variable star searches than are the larger amplitude RRab variables. Period distributions of the field RR Lyraes of type RRab and RRc are shown in figures 1.3 and 1.4 (Novikova 1988).

Large though these numbers may seem, only a small fraction of the Galaxy's RR Lyrae variables have been discovered. Suntzeff et al. (1991) estimated that the population of RR Lyrae variables in the halo of the Galaxy between 4 and 25 kpc from the galactic center is about 85 000. The total number of RR Lyraes in the Galaxy, including those in the galactic bulge and in the thick disk, is substantially larger still.

In 1944 Baade announced his discovery of two distinct stellar populations: the Population I stars of the disk and spiral arms, and the Population II stars of the halo and bulge. Most of the RR Lyrae stars of the Galaxy, including the vast majority of those found in globular star clusters, are metal-poor stars in the Milky Way's halo and nuclear bulge, and, initially, it appeared that the RR Lyrae stars could be safely consigned to population II. However, Preston (1959), among others, demonstrated that

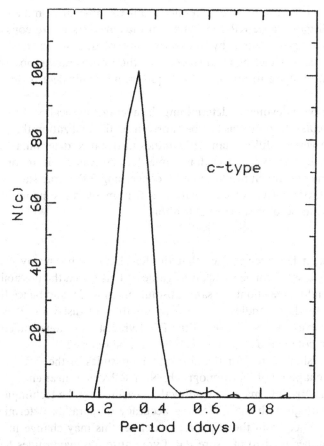

Figure 1.4 Histogram over period for field RRc stars in the Galaxy. While there may be a few RRc variables of greater or lesser period, most have periods between 0.18 and 0.5 days. From Novikova (1988).

the class of RR Lyrae stars actually includes a mixture of population types. He discovered that about 25 per cent of the field RRab stars in the solar neighborhood belong to an old disk population. In contrast to their Population II counterparts, these old disk RR Lyrae stars are only slightly to moderately deficient in heavy elements relative to the Sun, and have a rotation speed about the galactic center of about 200 km/s. These RR Lyrae stars are now believed to belong to the Galaxy's thick disk population (Gilmore, King, and van der Kruit 1990). Relatively metal-rich RR Lyrae stars have also been discovered in the galactic bulge, though most bulge RR Lyraes are metal-poor relative to the Sun.

RR Lyrae stars have been discovered beyond the limits of the Milky Way, first in the Magellanic Clouds and in the Milky Way's companion dwarf spheroidal systems, but recently in more distant Local Group galaxies, including the Andromeda Galaxy. The number of RR Lyrae stars discovered in these systems now exceeds 1500 and is sure to increase considerably .

1.2.2   *Mean physical parameters*

RR Lyraes are radially pulsating stars, but they are not unstable in the sense that novae or supernovae are unstable. Though they undergo oscillations, these oscillations are excursions around an equilibrium state which does not change significantly on the scale of a human lifetime. In discussing the RR Lyrae stars, it thus makes sense to consider that equilibrium state, as represented by measurements of the mean physical characteristics of the variables. Devising a satisfactory method of averaging observed properties over the pulsation cycle to arrive at the 'equilibrium' equivalent is often a difficult task.

Considerable effort has been devoted to determining the average physical parameters of stars of the RR Lyrae class. In part, this has been driven by the realization that RR Lyrae variables might be very useful as standard candles in distance determinations. Representative recent results are summarized in table 1.1. A caution is in order. Although this table summarizes much recent research concerning RR Lyrae stars, our knowledge of the physical parameters of these variables is in many ways incomplete. Each of the entries stands in need of some qualification.

Absolute magnitude

Since Bailey's early work, it has been known that the RR Lyrae stars in any given globular cluster show only a small range in mean brightness. This raised the possibility that all RR Lyrae stars might have about the same absolute magnitude, and hence that they might be excellent standard candles for the determination of distances. It is of course necessary that the absolute magnitudes of the RR Lyrae stars be determined in some fashion before their potential as standard candles can be realized.

The long and still unfinished search for the absolute magnitudes of the RR Lyrae variables is the subject of chapter 2 of this monograph. No RR Lyrae is near enough to have its distance, and thus absolute magnitude, directly measured by the technique of trigonometric parallax (though with the increasing accuracy of parallax determinations, such as may be obtained from the Hipparcos satellite, this may change in the future). Hence, the techniques used to measure RR Lyrae absolute magnitudes have been more indirect. Three basic approaches have been tried: statistical parallaxes, Baade–Wesselink solutions, and the determination by various independently calibrated

Table 1.1. *Properties of RR Lyrae stars*

Period	0.2–1.1 days
$\langle M_v \rangle$	+0.6 ± 0.2 (metal-poor stars)
$\langle T_e \rangle$	7400 K–6100 K
$\langle \log g \rangle$	2.5–3.0
[Fe/H]	0.0– −2.5
Mass	≈ 0.7 $M_\odot$
Radius	≈ 4–6 $R_\odot$

methods (Cepheid period–luminosity relations, main sequence fitting, etc.) of the distances to systems which contain RR Lyraes. The value $\langle M_v \rangle = +0.6 \pm 0.2$ in table 1.1 spans most recent determinations for metal-poor RR Lyrae stars, which generally range between +0.4 and +0.8. However, all RR Lyrae stars may not share exactly the same absolute magnitude and there has been much recent discussion as to whether RR Lyrae absolute magnitudes are a function of metallicity. There is substantial evidence that metal-rich RR Lyrae stars are less luminous than metal-poor RR Lyraes. The size of this difference in luminosity is a matter of controversy, but may amount to as much as a few tenths of a magnitude between stars differing by a factor of ten in heavy element abundance.

Surface temperature

As indicated in figure 1.2, stars of the RR Lyrae type are confined to well-defined limits in the HR diagram. These limits have been defined by observations of RR Lyrae stars both in the general field of the Galaxy and in globular star clusters. Average values of effective temperature and surface gravity have usually been calculated from application of model stellar atmospheres to multicolor photometry or from spectroscopy of those stars. Model atmospheres do not of course give a perfect representation of real stars and various systematic errors can occur in this process. For example, the hottest RR Lyrae stars (of type RRc) appear to have a mean effective temperature near 7400 K and the coolest RR Lyrae stars (of type RRab) appear to have a mean effective temperature near 6100 K, as indicated in table 1.1. However, these limits are subject to some uncertainty. Different color–temperature relations in the recent literature yield effective temperatures for the hottest and coolest RR Lyrae stars which differ by as much as 300 K from the values given above. It may be that those values are too high, perhaps by 100–200 K. The temperature width of the RR Lyrae instability strip, about 1300 K, is perhaps subject to less uncertainty. Similar uncertainties affect the derived values of surface gravity, $\langle \log g \rangle$, which, though certainly approximately correct, may be uncertain by a few tenths in the logarithm.

Chemical composition

When one measures the chemical composition of a star, either by multicolor photometry of its light, or by analysis of its spectral lines, one is actually measuring its surface, photospheric composition. In the case of relatively low-mass stars, such as those which become RR Lyrae stars, nucleosynthesis of significant amounts of elements heavier than carbon and oxygen is not expected during the lifetime of the star.

Thus, the measured abundance of heavy elements in the atmosphere of an RR Lyrae star is believed to reflect the abundance of heavy elements in the interstellar gas cloud from which that star formed. It is this property, together with their considerable age, which make RR Lyrae stars useful tracers of chemical history.

RR Lyrae stars are observed to differ considerably in chemical composition, particularly in their photospheric abundances of the elements heavier than helium ('metals' in astronomical parlance). As noted above, many RR Lyrae stars belonging to the old disk population of the Galaxy and some (but not all) of those found in the galactic bulge are only modestly deficient in heavy elements relative to the Sun. Some may be as metal-rich as the Sun and others perhaps even more so. On the other hand, those RR Lyraes found in the galactic halo can be very deficient in heavy elements.

To characterize the overall heavy element abundance of stars, astronomers often employ the [Fe/H] notation. In this notation, the ratio of iron to hydrogen in the photosphere of one star is related to that ratio in another star, usually the Sun: [Fe/H] = log(iron/hydrogen)$_*$ − log (iron/hydrogen)$_\odot$.

In this notation, the metal rich RR Lyraes are near, but usually somewhat below, [Fe/H] = 0.0, and the most metal-poor RR Lyraes are near [Fe/H] = −2.5, deficient in iron by a factor of 300 relative to the Sun. The [Fe/H] notation is a useful shorthand, which will be employed extensively in this book, but it must always be remembered

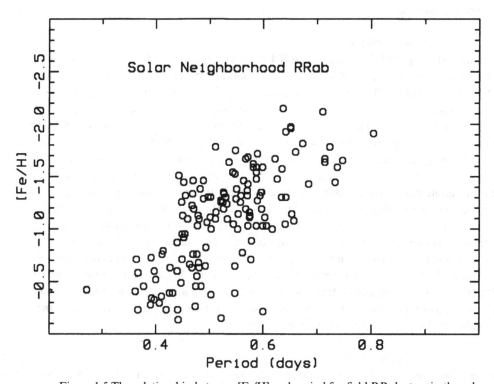

Figure 1.5 The relationship between [Fe/H] and period for field RRab stars in the solar neighborhood. The existence of a general correlation between period and metallicity for these stars is shown, with the more metal-rich RRab stars tending to have shorter periods. The metallicities are from diverse sources, but mainly from ΔS measurements. The metal abundance scale is that of Butler (1975).

that it is a simplification. The abundances of all heavy elements may not vary in lockstep, and a single number may not in reality suffice to describe the abundances of all elements heavier than helium.

The [Fe/H] values of solar neighborhood RRab and RRc stars are plotted as a function of period in figures 1.5 and 1.6. RR Lyrae stars which belong to the disk population generally have [Fe/H] > −1.0. Those in the outer halo of the Galaxy usually have −1.0 > [Fe/H] > −2.5, with an average metallicity near [Fe/H] = −1.6.

The metal abundances of RR Lyrae stars may also be expressed in terms of ΔS, a spectroscopic parameter devised by Preston (1959). Though based upon observations of the Balmer lines and the calcium K-line (see chapter 4), ΔS is well correlated with [Fe/H]. A star of $\Delta S = 0$ has a metal abundance similar to, but perhaps slightly less than, the Sun, whereas a star of $\Delta S = 10$ is very metal-poor at [Fe/H]≈ -1.8.

The abundance of the element helium is also believed to be important in determining the course of evolution and pulsational properties of an RR Lyrae star. Theory tells us that relatively small differences in helium abundance can have significant impacts upon the evolution and pulsation of otherwise similar horizontal branch stars. Unfortunately, the photospheric helium abundance of an RR Lyrae star cannot be measured spectroscopically with any accuracy. The uncertainty of spectroscopic helium abundances is too great to be useful in making comparisons with theoretical predictions. Instead, the abundance of helium in an RR Lyrae star is often determined

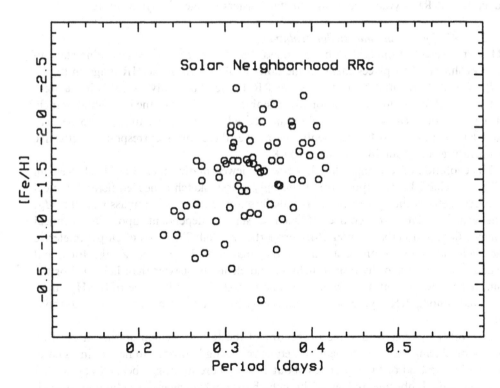

Figure 1.6 The relationship between [Fe/H] and period for field RRc stars in the solar neighborhood, after Kemper (1982). The metal abundance scale is again that of Butler (1975).

indirectly. For example, in the R-method the observed numbers of horizontal branch and red giant branch stars in a globular cluster are compared. Iben (1968) showed that the ratio of these numbers depended upon the cluster helium abundance. Such determinations indicate that stars which evolve to become RR Lyrae variables have initial helium abundances amounting to 20–30 percent of their total mass, with a value near 23 percent perhaps being most likely. The validity of these results are, however, dependent upon the adequacy of the theories employed, as Sweigart (1991) has reminded us.

It has been suggested that diffusion may change the photospheric chemical abundances of horizontal branch stars. Diffusion may occur, but at the time of writing no firm evidence of such diffusion has been produced for RR Lyrae variables.

Age

As discussed in §1.2.3, RR Lyrae stars are believed to be evolved low-mass stars. Their main sequence progenitors were therefore formed long ago. Age determinations for globular clusters which contain RR Lyrae stars help tell us how long ago. There is, however, still no unanimity on the ages of globular clusters, with age estimates for the same clusters differing by a few gigayears, depending upon which color–magnitude diagrams, distances, color–temperature relations, chemical compositions, and theoretical isochrones are employed. Values of 14–17 Gyr for the halo globular clusters are, perhaps, typical. Evidence from the Magellanic Clouds (chapter 6) suggests that RR Lyraes occur only in star clusters of age 12 Gyr or more.

1.2.3   RR Lyrae stars and stellar evolution

RR Lyrae stars are thought to be low-mass stars in the core helium burning stage of their evolution. This places them on the horizontal branch in the HR diagram (figure 1.7). Not all horizontal branch stars are RR Lyrae stars: only those within a well-defined instability strip are pulsationally unstable. At the level of the horizontal branch, at a luminosity near 40 $L_\odot$, the blue and red edges of the instability strip lie near $(B - V)_0 = 0.18$ and 0.40. As mentioned above, these limits correspond to effective temperatures of about 7400 K and 6100 K.

The evolution of low mass stars has been reviewed many times (Iben 1971; Renzini 1977; Castellani 1985; Caputo 1985) and only a brief sketch is needed here. A typical RR Lyrae star is thought to begin as a main sequence star with a mass near 0.8 $M_\odot$. The exact evolutionary course of such a star is dependent upon its chemical composition, and perhaps upon other properties as well. Theorists often parameterize the chemical composition of a star by the quantities X, Y, and Z, the fractional abundances by mass of hydrogen, helium, and elements heavier than helium. For the Sun, the heavy element abundance Z is about 0.02. The wide range of [Fe/H] values observed among RR Lyrae stars corresponds to a range from $Z = 10^{-4}$ to about 10^{-2}.

Like other low-mass stars, the progenitor of the RR Lyrae star spends most of its energy-producing lifetime on the main sequence, fusing hydrogen to helium in its core. After a long period of time, perhaps about 15 Gyr, core hydrogen becomes exhausted and the star climbs the red giant branch, burning hydrogen to helium in a shell surrounding a helium core. Temperatures in the helium core are not yet high enough for fusion of helium atoms to form still heavier elements, and the inert core collapses,

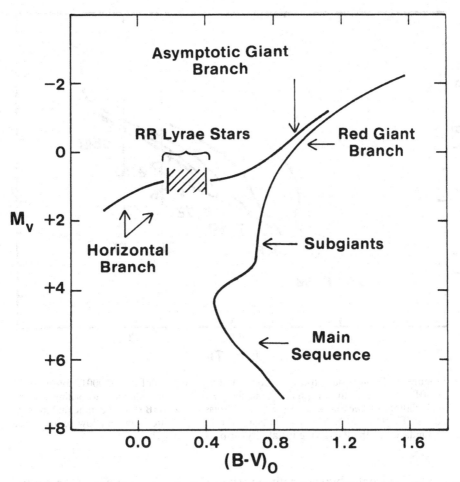

Figure 1.7 A schematic color–magnitude diagram for a typical globular cluster. The principal sequences are labeled. RR Lyrae stars are horizontal branch stars which happen to be located where the instability strip intersects the horizontal branch.

becoming electron degenerate. At the tip of the red giant branch the core temperature becomes high enough to initiate helium fusion. In a so-called helium flash, helium burning is ignited in the electron degenerate core via the triple alpha process. This removes the core degeneracy and the star proceeds to the zero-age horizontal branch (ZAHB) and the core helium burning stage of its existence. It is in this stage that, if the star falls within the bounds of the instability strip in the HR diagram, it pulsates as an RR Lyrae star. The RR Lyrae is a giant star at this point, with a radius 4–6 times that of the Sun, though both its radius and luminosity are much reduced from what they were at the tip of the red giant branch.

The location of a star on the ZAHB depends (at least) upon its total mass, its core mass, and its chemical composition. Stars in a single globular cluster are believed to have all started out on the main sequence with closely similar chemical compositions. Moreover, when they reach the ZAHB they are also expected to have similar core masses. Thus, in canonical evolutionary theory, the color sequence along the horizontal branch implies a mass sequence among the horizontal branch stars (figure 1.8).

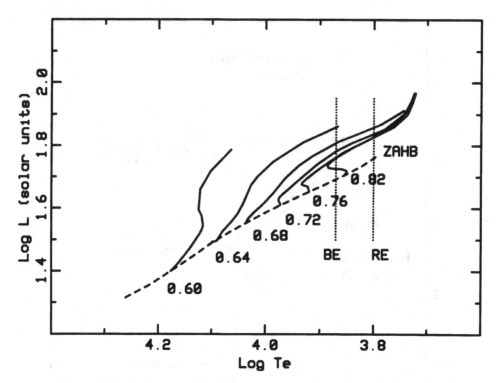

Figure 1.8 Evolutionary tracks of horizontal branch stars of $Z = 0.0001$, core mass = 0.5017 $M_\odot$, and initial (main sequence) $Y = 0.23$ are shown, according to the calculations of Lee and Demarque (1990). Masses of ZAHB stars are indicated in solar units. The dotted vertical lines indicate the approximate high and low temperature boundaries of the RR Lyrae instability strip.

Because the horizontal branch lifetime is short compared to the age of the cluster, stars which are found at any one time on the horizontal branch are believed to have had nearly identical main sequence masses. Thus, the spread in color among horizontal branch stars implies that different horizontal branch stars have experienced different amounts of mass loss before reaching the ZAHB, with losses of 0.1 $M_\odot$ perhaps being typical. However, the determination of RR Lyrae masses from theoretical isochrones, such as those in figure 1.8, is complicated by the circumstance that not all RR Lyrae stars are ZAHB stars. Some may have evolved into the instability strip from ZAHB positions outside the RR Lyrae zone. Nevertheless, stellar evolutionary masses near 0.7 $M_\odot$ are typically obtained for RR Lyrae variables. These evolutionary masses are especially important because, although masses of stars in appropriate binary star systems can be directly measured, no RR Lyrae star has been discovered in a binary system suitable for mass determination.

RR Lyrae masses must instead be determined either from stellar pulsation or stellar evolution theory. RR Lyrae masses determined from stellar pulsation theory can be compared with those determined from stellar evolution theory. Particularly useful for this purpose are RRd stars, RR Lyraes which are pulsating simultaneously in the fundamental and first overtone radial modes (§5.3) Unfortunately, uncertainties in the chemical compositions and opacities of these stars presently introduce an uncertainty of at least 0.1 $M_\odot$ in the masses obtained from pulsation theory. Recent values of 0.65 to 0.8 $M_\odot$ for the masses of the RRd stars are higher than those obtained a few years earlier, but nonetheless remain in at least rough agreement with the masses given by stellar evolution theory.

Mass is very unequally distributed throughout the RR Lyrae star. The dense core, with a radius of about 5 Earth radii, has a mass of about 0.5 $M_\odot$. On the other hand, the outer 50 percent of the star's radius is of very low density and contains little mass, on the order of 0.001 $M_\odot$.

Though we call the horizontal branch stage the core helium burning phase, a horizontal branch star actually has two sources of nuclear energy production. Helium is fusing to carbon and oxygen in the core, but hydrogen is also fusing to helium in a shell surrounding the helium core. Eventually, after about 10^8 years, the central helium is depleted. As core helium exhaustion approaches, the star leaves the horizontal branch. It swells and cools once again, ascending the asymptotic red giant branch and deriving energy from a hydrogen burning shell and a helium burning shell. Ultimately, the asymptotic red giant runs out of useable nuclear fuel. Its central temperature never becomes high enough for fusion of still heavier elements to begin in its carbon and oxygen core. Probably after expelling its outer gaseous envelope as a planetary nebula, the star continues to shine feebly as a white dwarf. Without nuclear reactions to power it, the white dwarf gradually radiates away its internal heat energy, though at a slow rate because of its very low luminosity.

1.2.4   The pulsation cycle: observations

Most observed properties of RR Lyrae stars vary during the course of a single pulsation as the star undergoes coincident changes in effective temperature and radius. It is the cyclic variations in radial velocity that demonstrate most clearly that the star is, in fact, pulsating. The luminosity of the star is affected by both the changes in radius (R) and in effective temperature (T_e). Figure 1.9 shows that the light amplitude of an RR Lyrae star is a strong function of wavelength. The star is bluer at maximum than at minimum and the amplitude in B (at about 4400 Å) exceeds that in V (about 5500 Å), while both the B and V amplitudes are much greater than the infrared K amplitude (2.2 μm). This dependency is readily understood if the star is considered to behave very roughly as a blackbody radiator. Visible wavelengths fall near the peak of the Planck function for an RR Lyrae star. Thus, visible (as well as bolometric) fluxes scale as $R^2 T_e^4$. In contrast, the K passband lies on the Raleigh–Jeans tail of the Planck function and the K-band flux scales as $R^2 T_e^{1.6}$ (Jameson 1986). As a consequence of these scaling relations, the luminosity changes in the B and V passbands mainly reflect the changes in effective temperature of the star during the pulsation cycle, whereas the infrared amplitude is more dependent upon the star's changes in radius.

Multicolor photometry has now been obtained over the complete light cycle for many RR Lyrae stars. A typical RRab star might have an effective temperature of 7500 K at visual light maximum, falling to 6100 K at minimum light. However, not all RRab stars show the same range of variation in surface temperature: there are significant star-to-star differences, especially near maximum light. As a rule, RRc stars have higher effective temperatures than RRab stars, both in the mean and at minimum

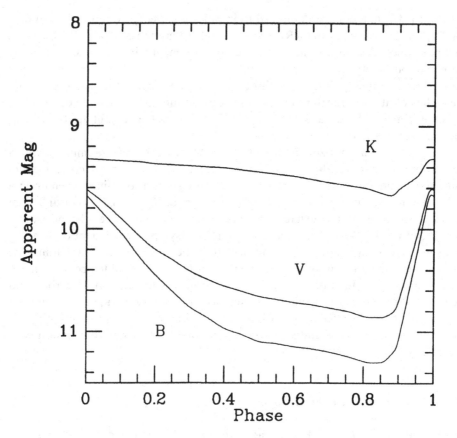

Figure 1.9 This figure compares the *B* (blue light), *V* (green-yellow light), and *K* (infrared) lightcurves of the RRab star RS Boo. The amplitude decreases as the wavelength increases. Adapted from Jones et al. (1988b).

light. Thus, the cooler portion of the RR Lyrae instability strip tends to be populated by RRab variables, while the hotter portion is occupied by RRc variables.

The line spectrum of an RR Lyrae star also undergoes a cyclic change during the pulsation cycle (figure 1.10). Spectroscopic observations of the brighter field RR Lyraes have a long history, but spectral classification of these variables has been complicated by their wide range in metal abundance. Normal MK classification criteria are not usually applicable to metal deficient stars, and thus unambiguous spectral typing is difficult for the many metal-poor RR Lyraes. Preston (1959), for example, found that he often obtained very different spectral types for RR Lyrae stars depending upon whether he classified them according to the strength of the K-line of CaII (λ3933) or the hydrogen Balmer lines. Preston was able to put these differences to good use in his ΔS method of determining the metal abundances of these stars.

Effective temperature is the main determinant of spectral type among the stars of Population I. Given their range in metal abundance, can one define a similar temperature dependent spectral type for the RR Lyrae variables? To a degree, the answer is yes. The star-to-star range in photospheric hydrogen abundance among the RR Lyrae stars is much less than for the heavy elements. Thus, if one seeks to define a

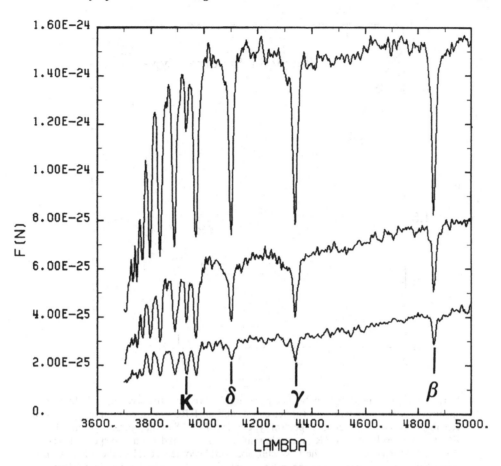

Figure 1.10 The change in the spectrum of the intermediate metallicity RRab star RW Dra from maximum light (top) to minimum light (bottom). The K-line of CaII and the Balmer δ, γ, and β lines are identified. The spectra have been slightly shifted in flux to avoid overlapping.

spectral type for an RR Lyrae star which reflects principally effective temperature, a type based upon the strengths of the hydrogen Balmer lines is superior to one based upon the metal lines. Figure 1.11 illustrates the changes in hydrogen spectral type through the light cycle for typical RRab and RRc stars.

Because RRab variables can show emission lines during rising light, and because the effective surface gravity changes rapidly during the rise, caution must still be used in interpreting the hydrogen spectral type during those phases. The occurrence of this emission, together with doubling of some spectral lines at certain phases during rising light, is a strong indication that shock waves are present in the atmospheres of these stars.

The pulsation of an RR Lyrae star induces a periodic variation in the radial velocity of the star. The radial velocity curve roughly reflects the lightcurve, with minimum radial velocity coming at visible light maximum for RRab variables (figure 1.12). For RRc stars the minimum radial velocity is reached about 0.1 of a cycle after maximum light. The radial velocity amplitude, like the light amplitude, differs from star-to-star,

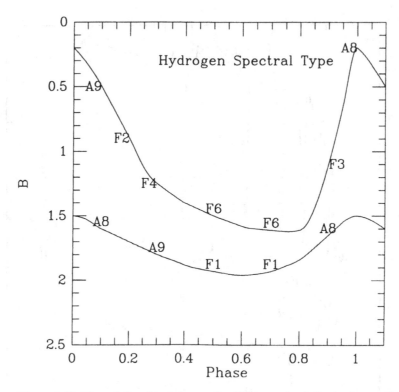

Figure 1.11 The relationship between the lightcurve and the hydrogen Balmer line spectral type for an RRab star (upper curve) and an RRc star (lower curve). A typical RRab star has a hydrogen line spectral type near A7 or A8 at maximum light and F5 or F6 at minimum light. An RRc variable may also have a hydrogen line spectral type of A7 or A8 at maximum, but the RRc star does not become as cool as the RRab star at minimum light, so that its hydrogen line spectral type rarely gets later than F1 or F2.

but amplitudes of 60–70 km/s are typical for RRab's and 30–40 km/s for RRc's. The observed radial velocity also depends upon which spectral lines are measured. For much of the pulsation cycle, instantaneous velocities based upon measurements of the strong Balmer lines are, for example, systematically different from those based upon weak metallic lines – the hydrogen lines generally giving a larger radial velocity amplitude. This presumably reflects the circumstance that lines of different elements and ionization states can arise from different levels within the moving stellar atmosphere. Discontinuities can be seen in radial velocities determined from measures of strong absorption lines in the spectra of some RRab stars. These occur during rising light and are attributed to shock wave phenomena.

The cyclic change in radius associated with the pulsation can be determined by integrating over the weak-line radial velocity curve, though a correction factor must be included to account for projection effects and limb darkening. The calculation of the actual physical and angular radius of the star is of course a much more complicated procedure than the determination of the change in radius alone. Several recent attempts have been made to apply the Baade–Wesselink method to RR Lyrae stars to obtain these radii (figure 1.13). The difference between the maximum and minimum radius of an RR Lyrae star is on the order of 15 percent of the mean radius.

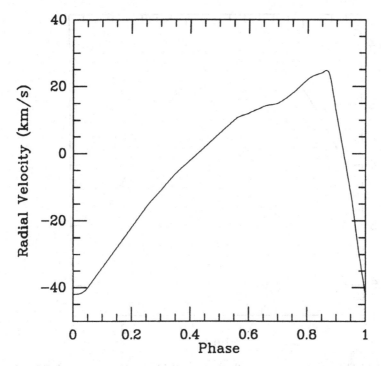

Figure 1.12 The radial velocity curve of RS Boo, adapted from Jones et al. (1988b). This curve is almost a mirror inverse of the lightcurves shown in figure 1.9. The radial velocity curve shown is based upon observations of metallic lines. The similar curve based upon observations of the Balmer lines will differ, especially during rising light.

1.2.5 The pulsation mechanism

Excellent reviews of the theory of stellar pulsation (Cox 1974; King and Cox 1968), including a recent monograph (Cox 1980), make superfluous any detailed treatment of the subject in this volume. Nevertheless, in subsequent chapters the confrontation of pulsation theory with observation will be explored and a brief reprise of the pulsation mechanism in RR Lyrae stars is therefore in order.

August Ritter as early as 1873 suggested that some stars might be variable because of radial pulsations. He went on to calculate the relation which bears his name for the slowest pulsation period of a homogeneous star:

$$P\sqrt{\rho} = Q,$$

where P is the period, ρ is the density of the star, and Q is a constant, usually called the pulsation constant. This equation is also commonly known as the pulsation equation. If P is measured in days, and ρ in units of the mean solar density, then Q is typically 0.04 for RR Lyrae stars of type ab.

Despite Ritter's calculations, and those of several others who suggested that pulsations of different types might play a role in stellar variability, at the turn of the century the most widely accepted explanation for Cepheid and RR Lyrae variability was that stars of that type were ill-understood spectroscopic binaries. Plummer (1913) and, especially, Shapley (1914) argued for the pulsation hypothesis, but it was

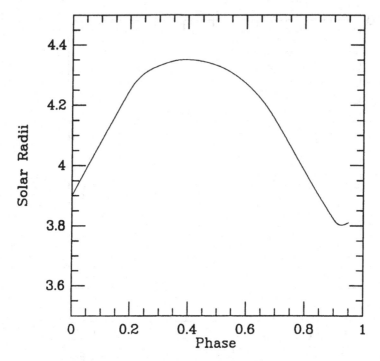

Figure 1.13 The change in radius with phase for RS Boo, after Jones et al. (1988b). The phase scale is the same as in figure 1.12.

Eddington who, in a series of papers beginning in 1917, developed the mathematical foundation of radial pulsation theory.

In his influential 1926 book *The Internal Constitution of the Stars,* Eddington devoted a chapter to stellar pulsation. He redeveloped Ritter's relation for a more realistic model star and went on to consider how pulsation might be maintained. He found that ordinarily a radial pulsation, once initiated, would decay on a timescale of about 8000 years. This was such a short time compared to reasonable stellar lifetimes that one would expect to find very few pulsating stars in the Galaxy. Later calculations confirmed Eddington's result, giving decay times which were even shorter. The problem, then, was to discover a mechanism which would allow stars within the instability strip to keep pulsating, but which would be ineffective for stars outside the instability strip.

What was needed, Eddington concluded, was a mechanism that would allow an RR Lyrae star or Cepheid to act as a thermodynamic heat engine, adding heat in the high temperature part of the pulsation cycle and releasing it in the low temperature part. To do this, the star must have a 'valve' to regulate the flow of energy.

Eddington first sought this valve in a variable rate of nuclear energy generation near the center of the star. We now know, however, that Cepheids and RR Lyraes have a strong central mass concentration and that the pulsation phenomena are restricted to the outer envelope which contains a small percentage of the mass. Epstein's (1950) calculations for centrally condensed stars showed that the radial variation of the surface layers is roughly a million times larger than that of the central regions where nuclear energy is produced. The pulsations of the variables have essentially no direct effect upon conditions deep in the interior. The rate of energy generation is constant over the pulsation cycle, and it is instead in the envelope that the valve mechanism must be sought.

Eddington himself contemplated a second type of valve:

> We now consider another position of the 'valve' – fantastic in an ordinary engine but not necessarily so in the star. Suppose that the cylinder of the engine leaks heat and that the leakage is made good by a steady supply of heat. The ordinary method of setting the engine going is to vary the *supply* of heat, increasing it during compression and diminishing it during expansion. That is the first alternative we considered. But it would come to the same thing if we varied the *leak*, stopping the leak during compression and increasing it during expansion. To apply this method we must make the star more heat-tight when compressed than when expanded; in other words, the opacity must increase with compression.

It is this second valve mechanism which drives the pulsations of RR Lyraes and Cepheids, but it was some time after Eddington's suggestion before the origin of the opacity changes could be elucidated. When *The Internal Constitution of the Stars* was published, it was still believed that hydrogen and helium were relatively minor constituents of stars. Within a few years, however, it was recognized that the two light elements were the main stuff of stars. Eddington later considered whether the ionization of hydrogen in the outer layers of stars might play a role in maintaining pulsations, but the realization of the importance of ionization zones to pulsation is attributable to Zhevakin (1953) and, later, Cox and Whitney (1958). They showed that the region in which helium becomes doubly ionized could serve as the site of Eddington's valve.

The overall opacity of a region in a star can be characterized by the Rosseland mean opacity, κ. The temperature and density dependency of the Rosseland mean opacity can be described by the relation

$$\kappa = \kappa_0 \rho^n T^{-s},$$

where κ_0 is a constant, ρ is density, and T is temperature. For the gas in the stellar envelope where no important element is undergoing ionization, n is about 1 and s about 3.5. As T increases, the opacity decreases. The leakage of energy would therefore increase during compression and the gas cannot serve as Eddington's valve. However, in a region where an abundant element is being ionized, s can become small or even negative with the result that the gas can become most opaque near maximum compression (King and Cox 1968). This is, of course, just what is needed to create Eddington's valve. This valve mechanism has become known as the κ-mechanism.

There is a second way in which the ionization zones in an RR Lyrae star can act to produce Eddington's valve. Energy which would ordinarily go into raising the temperature of the gas can instead go into ionizing material. The layer will therefore 'tend to absorb heat during compression, leading to a pressure maximum which comes after minimum volume and hence provide a driving [force] for pulsations' (King and Cox 1968). This effect has been called the γ-mechanism.

For RR Lyrae stars the zone where helium is being doubly ionized is believed to play the most important role in driving the pulsations, but the zone where hydrogen is ionized is also important (Christy 1966). The temperature of the helium second ionization zone is 3×10^4 to 6×10^4 K, whereas the hydrogen ionization zone occurs at temperatures less than 3×10^4 K. These temperatures are found very high in the envelope of an RR Lyrae star. Even the high temperature edge to the helium second ionization zone occurs at radii more than 90 percent of the distance from the center of the star to the surface.

In addition to the κ and γ-mechanisms, it is possible that there are other mechanisms which contribute to driving the pulsation. Cox (1985) lists six other possible excitation mechanisms for pulsating stars. It appears likely, however, that most of these are unimportant for the RR Lyrae stars.

There is a second part of the problem. We have noted that RR Lyrae stars are observed to occupy a specific region of the HR diagram. Horizontal branch stars which are too hot or too cool at the surface do not pulsate. Any theory describing RR Lyrae pulsation must not only explain why stars within the instability strip pulsate, but why stars outside the strip do not.

The precise high and low temperature boundaries to the RR Lyrae instability strip are difficult to predict theoretically because such a prediction requires an understanding of the role of the complicated process of convection in RR Lyrae pulsation (see chapter 3). That is particularly true for the low temperature (red) edge to the RR Lyrae instability strip. It is believed that horizontal branch stars cooler than the red edge do not pulsate because the onset of convection disrupts the pulsation mechanism. The location of the hot (blue) edge to the instability strip is easier to calculate because the envelopes of stars at that temperature can be more easily described by models in which radiation is the dominant means of energy transport. Pulsation halts at the blue edge because the helium second ionization zone is too high in the atmosphere of the star. The density of material at that high level in the atmosphere is too low for the ionization zone to effectively dam the flow of energy and serve as an adequate valve to drive the pulsations. It was at one time believed that purely radiative models would suffice for predicting the location of the blue edge to the RR Lyrae instability strip, but more recent calculations by Stellingwerf (1984) have suggested that convection may also play some role in determining the blue edge location.

1.2.6   *Pulsation modes*

When the RR Lyrae stars in an RR Lyrae-rich globular cluster are plotted in the period–amplitude diagram, they divide into two distinct groups, as shown in Figure 1.14. The low-amplitude, short-period group in this diagram is comprised of RR Lyrae stars of Bailey type c. In the longer period group, the higher amplitude RR Lyraes belong to Bailey type a, while the lower amplitude RR Lyraes are of Bailey type b. There is no sharp distinction between the type a and type b RR Lyraes in this diagram, instead there is a smooth transition from type a to type b. For this reason it is clear that the distinction between Bailey type a and b RR Lyraes is less significant than the distinction between those two types and the type c RR Lyrae stars. Bailey himself suggested as much in his original definitions of the RR Lyrae types. It is more generally useful to consider RR Lyrae variables as divided into just two basic types: the RRab and RRc variables.

Martin Schwarzschild (1940) argued that RRab stars in the globular cluster M3 are pulsating in the fundamental mode, whereas the RRc stars pulsate in the first overtone

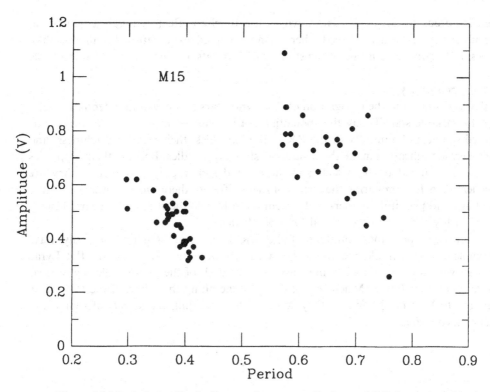

Figure 1.14 Period–amplitude diagram for some well-observed RR Lyrae variables in the globular cluster M15. The RR Lyraes divide into two groups: the longer period RRab variables and the shorter period RRc variables (after Bingham et al. 1984).

mode. This suggestion has since been confirmed by detailed computations of RR Lyrae pulsations and generalized to all RRab and RRc pulsators. An RRc variable thus has one pulsational node in its interior and would have a smaller pulsation constant Q than an RRab star of equal density. For a typical RR Lyrae, stellar pulsation theory predicts, and observations of double mode RR Lyraes have confirmed, that the first overtone pulsation period is about 3/4ths as long as the fundamental mode period. As noted above, RRc variables have been observed to have higher mean surface temperatures than RRab variables. Whether the dividing line in effective temperature between RRc and RRab pulsation is the same for all globular clusters is still a matter of contention. In at least some clusters, there may exist a temperature zone in which both RRc and RRab pulsation exist.

It has sometimes been suggested that a few RR Lyrae stars might pulsate in the second overtone mode (e.g., Demers and Wehlau 1977). Nevertheless, the existence of second overtone pulsators has yet to be firmly established, and theory is at best equivocal on whether second overtone RR Lyraes should exist (Stellingwerf et al. 1987; Stothers 1987). A number of RR Lyrae stars have been found, however, that show mixed mode behavior, pulsating simultaneously in the first overtone and fundamental modes. These double mode RR Lyraes, often called RRd stars, have ratios of first overtone mode to fundamental mode periods of about 0.74 to 0.75.

Some RRab stars show a modulation of the primary light cycle on a scale of tens or even hundreds of days. This is known as the Blazhko effect. Though a number of hypotheses have been advanced, there is no accepted explanation for the Blazhko cycles. RR Lyrae itself is one of the stars which exhibits a secondary cycle of this type.

1.3   Summary

Bailey's efforts led to the recognition of RR Lyrae stars as a new and potentially useful type of variable star. Today that potential can be realized in a variety of ways. The limited spread in luminosity of the RR Lyrae variables, their relative brightness, and their easy identification make them valuable standard candles. Because their properties can be determined by both stellar evolution and stellar pulsation theory, they are excellent stars for comparing the results of these different theoretical approaches. Their great age and plentiful numbers make them important to the study of the early history of the Milky Way and other Local Group galaxies.

Nonetheless, our understanding of the RR Lyrae stars remains in many ways incomplete. For example, we do not know the absolute magnitudes of the RR Lyrae stars as well as we would wish, nor have we solved all of the puzzles relating to their pulsation and evolution. Much remains to be done along these lines if the RR Lyrae stars are to become the probes they could be for revealing the secrets of stellar and galactic evolution.

2

The absolute magnitude of the RR lyrae stars

Several factors combine to make the RR Lyrae stars potentially good standard candles. First, though they are not as bright as the Cepheid variable stars, they are luminous giant stars which can be detected to considerable distances. Second, there are many more RR Lyrae stars known in the Galaxy than there are Cepheids, and the RR Lyrae stars occur in old systems which do not contain Population I distance indicators such as the classical Cepheids. Third, the range in absolute magnitude of the RR Lyrae stars appears to be quite restricted. Fourth, RR Lyrae stars, like Cepheids, have the virtue that they can be easily identified from their light variations. For these reasons the absolute magnitudes of the RR Lyrae variables have long been sought. However, though the approximate absolute magnitude of the RR Lyrae stars is known, pinning down this value to the desired accuracy (within ± 0.1 mag) has proved a difficult and still unfinished task.

As noted in chapter 1, trigonometric annual parallaxes of the RR Lyrae stars are still too uncertain to usefully constrain their absolute magnitude. This may change in the future, but for now the absolute magnitude must be fixed by more indirect means. There have been three main approaches to measuring RR Lyrae star luminosities. First, their mean absolute magnitude may be determined by the technique of statistical parallaxes. Second, absolute magnitudes of individual RR Lyraes can be determined by various versions of the Baade–Wesselink method (Wesselink 1946). Finally, absolute magnitudes of RR Lyrae stars in various stellar systems can be determined by reference to other stars of known absolute magnitude in the same system. This last approach has been applied mainly through fitting the main sequences of globular clusters to the local subdwarf main sequence, or by independently determining the distance to the Magellanic Clouds and thereby fixing the absolute magnitudes of the RR Lyrae stars they contain.

2.1 Background

As we have seen, Bailey had found that the apparent, and hence the absolute, magnitudes of the RR Lyrae variables within a single globular cluster showed a small dispersion. Moreover, in each cluster these variables seemed to be about 1.5 or 2 magnitudes fainter than the brightest cluster members. It was a plausible initial assumption that all RR Lyrae variables, no matter whether they occurred in or out of a globular cluster, had essentially the same absolute magnitude.

This property, and their occurrence in many of the globular star clusters, made the RR Lyrae variables of potential great use to Harlow Shapley in his ground-breaking investigations of the distances of the globular clusters. Leavitt (1908, Pickering 1912) had discovered that Cepheids in the Small Magellanic Cloud obey a period–luminosity relation. Shapley (1918) had determined the zero-point of this relation by a statistical parallax analysis of only 11 Cepheids. To extend this period–luminosity relation to the RR Lyrae variables, Shapley took note of the fact that some globular clusters contained both Cepheids and RR Lyrae stars. Baade's discovery of stellar Populations I and II was almost three decades in the future, and, given the data at his disposal, it was reasonable for Shapley to conclude that all Cepheids, including those in globular clusters, obey the same period–luminosity relation. By applying his calibration of the period–luminosity relation to the Cepheids in globular clusters, Shapley could determine not only the distances to those globular clusters, but also the absolute magnitudes of the RR Lyrae stars they contained. Shapley found that the RR Lyrae variables had a median absolute photographic magnitude of −0.23, a value about a magnitude brighter than present estimates.

Fernie (1969) has reviewed the development of the period–luminosity relation in the 50 years after Shapley's original calibration. Until Baade's 1952 announcement that there were two period–luminosity relations, most results more or less supported Shapley's calibration. In his 1930 book *Star Clusters*, Shapley gave a slightly fainter absolute magnitude for the RR Lyraes, $M_{pg} = 0$ and $M_v = -0.1$, results still quoted by Payne-Gaposchkin and Gaposchkin in their 1938 monograph *Variable Stars*. Nevertheless, there were a few indications that an absolute magnitude near 0.0 might be too bright. An analysis of RR Lyrae statistical parallaxes by Bok and Boyd (1933) derived an absolute photographic magnitude of $+0.08 \pm 0.15$, but Fletcher (1934) suggested a revised weighting scheme for their results which would give an absolute magnitude of $+0.44$. However, not until the 1950s would the value of 0.0 be seriously challenged.

In 1953, Pavlovskaya presented a new statistical parallax analysis for RR Lyrae stars which gave $M_{pg} = +0.5$, corresponding to $M_v = +0.2$. At about the same time, Sandage, Arp, and others began to obtain deep, photoelectrically calibrated, color-magnitude diagrams for globular clusters. Their early attempts to determine the distances to the globular clusters by main sequence fitting suggested that the RR Lyrae stars might form an inhomogeneous group, with RR Lyraes in different globular clusters having different absolute magnitudes.

The progress made in the1950s inspired numerous fresh attempts to fix the absolute magnitude of the RR Lyrae variables. The remainder of this chapter will focus upon the more recent of these attempts. The problem has proved more intractable than one might have hoped, partly because the RR Lyrae absolute magnitudes must be known very accurately (to within ± 0.1 mag) if the RR Lyraes are to be most effectively used as standard candles. Consideration of the absolute magnitude of the RR Lyrae stars has also been extended by the arrival of new observational techniques, of which photometry of RR Lyrae stars in the infrared is perhaps the most significant. Much recent work on the absolute magnitudes of RR Lyrae stars has been motivated by the need to obtain accurate distances to the globular star clusters in order to determine their ages.

2.2 Defining mean magnitude

Before we can discuss the absolute magnitude of an RR Lyrae star it is necessary to consider how to average the changing luminosity of the star over the light cycle. The simplest mean magnitude is

$$m = \tfrac{1}{2}(m_{\text{max}} + m_{\text{min}}),$$

where m_{max} and m_{min} are the observed magnitudes at maximum and minimum brightness, respectively. This magnitude differs, however, from the magnitude at mean intensity or the arithmetic mean of magnitudes averaged over the light cycle (Jameson 1986). For V magnitudes, Clube and Dawe (1980) give the relationship between these quantities as typically

$$\langle V \rangle = V + 0.07 = \langle V \rangle_t + 0.02,$$

where $\langle V \rangle$ is the magnitude at mean intensity, V is the magnitude midway between maximum and minimum , and $\langle V \rangle_t$ is the arithmetic mean of V magnitudes over the light cycle. The difference between these alternative types of mean magnitude is actually a function of lightcurve shape, however.

Different investigations have employed different definitions of mean magnitude and, at the 0.1 magnitude level, these differences must be taken into account in comparing their results. Most recent investigations have defined the mean magnitude by integrating the lightcurves in intensity units to derive the intensity mean magnitude. Since the energy generated by nuclear reactions in the core of the star is unaffected by the star's pulsations, this intensity weighted mean magnitude should be the same as would be observed were the star not pulsating.

A word is in order, too, about the different magnitude systems commonly encountered in the literature on RR Lyrae absolute magnitudes. In work since the 1950s, the absolute magnitude in the Johnson V-passband is most often used. This is a useful approach, not only because V data are available for many RR Lyraes, but because the horizontal branch in the neighborhood of the instability strip is nearly flat in V, as its name implies. For RR Lyraes in a given globular cluster, the trend of $\langle V \rangle$ with either period or color is usually small. This is also true of mean magnitudes measured in the Johnson B-passband or the older m_{pg} photographic system. However, the hottest RRc variables have values of $\langle B{-}V \rangle$ about 0.2 smaller than the coolest RRab variables. As a consequence, $\langle B \rangle$ is usually not quite so constant across the instability strip as is $\langle V \rangle$ – i.e., the horizontal branch is not so horizontal in B as it is in V. In the infrared, as will be discussed later, there is a significant dependence of mean magnitude upon period and the horizontal branch cannot be regarded as 'horizontal'.

2.3 RR Lyrae evolution and absolute magnitudes

There is one further item to be considered before proceeding to the discussion of recent determinations of RR Lyrae absolute magnitudes. This is the issue of the evolution of the RR Lyrae stars. After helium flash, evolving low-mass stars move to their positions on the zero-age horizontal branch (ZAHB). Stellar evolution theory tells us that, as these ZAHB stars gradually evolve, they will become brighter (figure 1.8). The difference between the luminosity of a ZAHB RR Lyrae and an RR Lyrae nearing the end of its horizontal branch life can amount to a few tenths of a magnitude, even if all else is equal. Therefore, even if we consider two RR Lyrae stars in the same globular cluster, we may see them at different stages during their horizontal branch lives, and therefore at different luminosities. Sandage (1990a) investigated the spread in apparent magnitude among RR Lyrae stars in nine globular clusters, finding evidence for a spread in magnitude which may be attributable to evolution.

RR Lyrae evolution will affect the different techniques for determining RR Lyrae absolute magnitudes in different ways. Statistical parallax results, which apply to a group of RR Lyrae stars, presumably indicate the luminosity of some average evolutionary state for field RR Lyrae stars. Main sequence fitting techniques can in principle provide information both upon the average luminosity and the spread in luminosity among the RR Lyrae stars in a particular globular cluster. Baade–Wesselink solutions are derived for individual RR Lyrae stars, and thus reflect the particular evolutionary state of the star under consideration. In two stellar systems of equal chemical composition, the typical RR Lyrae star may be more evolved in one than in the other, resulting in different mean RR Lyrae absolute magnitudes. This difference would presumably be just one aspect of an overall difference in horizontal branch morphology between the two systems.

In the end, if one is to use RR Lyrae stars to determine the distance to a system, the evolutionary state of the RR Lyrae stars in that system must be evaluated if the most accurate results are to be obtained.

2.4   Statistical parallax solutions

Statistical parallax solutions for the absolute magnitudes of RR Lyrae stars have a long history and many different solutions have been attempted. Since the 1960s, statistical parallax solutions have yielded values of $\langle M_v \rangle$ between $+0.3$ and $+1.0$. Van Herk (1965) obtained a mean magnitude of $\langle M_v \rangle = +0.6$, Woolley et al. (1965) obtained $\langle M_v \rangle = +0.52$, and Hemenway (1975) obtained $\langle M_v \rangle = +0.49$. These investigations used the traditional least squares approach to obtaining the statistical parallax solution, an approach which has been criticized by later workers as requiring too many assumptions. Heck and Lakaye (1978) and Clube and Dawe (1980) introduced maximum likelihood techniques into the statistical parallax analysis. The results of their analyses were, however, rather discrepant. Heck and Lakaye found $\langle M_v \rangle = +0.3$ at $\Delta S \approx 5$, with evidence that luminosity decreased with decreasing metallicity. Clube and Dawe, on the other hand, found $\langle M_v \rangle = +1.0 \pm 0.25$ for metal-poor RR Lyraes. These applications of the maximum likelihood technique have subsequently been criticised on various points (Hawley et al. 1986; Strugnell, Reid, and Murray 1986).

More recently, two groups have again tackled the question of the statistical parallaxes of the RR Lyrae stars (Strugnell, Reid, and Murray 1986; Hawley et al. 1986; Barnes and Hawley 1986). These investigators have employed the same sets of proper motions (Wan, Mao, and Ji 1981) and radial velocities. Although their methods for correcting for interstellar reddening and deriving mean magnitudes are somewhat different, both used intensity mean magnitudes and similar maximum likelihood methods. Strugnell et al. obtained $\langle M_v \rangle = 0.75 \pm 0.2$ for metal-poor RR Lyraes with $\Delta S = 5$–9 (a mean [Fe/H] near -1.3), with only an insignificant indication that luminosity increased as metallicity decreased. For a sample of 142 RRab stars covering a range of metallicity, Hawley et al. (1986) obtained $\langle M_v \rangle = +0.76 \pm 0.14$. They found no significant evidence for a trend with metallicity, but only a fairly strong trend would have been certainly detected by them. Barnes and Hawley (1986) later slightly

Table 2.1. *Mean visual absolute magnitudes for RR Lyrae stars*

Sample	$\langle M_v \rangle$	Method	Source
RRab $5 \leq \Delta S \leq 9$	$+0.75 \pm 0.2$	Stat. π	Strugnell et al. 1986
RRab	$+0.68 \pm 0.14$	Stat. π	Barnes and Hawley 1986
RRab	$+0.80 \pm 0.14$	Stat. π	Barnes and Hawley, as adjusted by Carney et al. 1992
RR	$+0.62 \pm 0.12$	Stat. π	Zhao 1988
RR(M5)	$+0.86 \pm 0.12$	ms-fitting	Jones et al. 1988b
RR(ZAHB)	0.39[Fe/H] + 1.32	ms-fitting	Buonanno et al. 1990
RR(field)	0.16[Fe/H] + 1.02	BW	Jones et al. 1992a
RR(field)	0.20[Fe/H] + 1.04	BW	Cacciari et al. 1992
RR	0.30[Fe/H] + 0.94	Blue edge	Sandage 1993b
RR(LMC)	0.15[Fe/H] + 0.73	LMC dist.	Walker 1992c
RR(IC1613)	$+0.46 \pm 0.15$	Ceph.	Saha et al. 1992
RR(LMC)	$+0.44$	Puls. theor.	Simon and Clement 1993
RR	0.51[Fe/H] + 1.58	Astrometry	Rees 1993

corrected this result to $\langle M_v \rangle = 0.68 \pm 0.14$. The mean metallicity of the Hawley et al. RRab sample is about [Fe/H] $= -1$. For 17 RRc variables, Hawley et al. obtained $\langle M_v \rangle = +1.09 \pm 0.38$. Considering the large uncertainty in the solution for the small RRc sample, the difference in absolute magnitude between the RRab and RRc stars is not significant. Carney et al. (1992) found that the Barnes and Hawley mean magnitude for RRab stars becomes $\langle M_v \rangle = +0.80 \pm 0.14$ when converted to the reddening scale used by Strugnell et al. Zhao (1988), also applying a maximum likelihood method, obtained the rather similar value, $\langle M_v \rangle = +0.62 \pm 0.12$.

Recent statistical parallax results are in good agreement, as they should be since they are derived from similar data and methods. They suggest a relatively small dispersion in absolute magnitude among the RR Lyrae stars. However, when the samples used in the statistical parallax solutions are divided into subgroups by metallicity, the errors associated with the results for each subgroup are too large to exclude a modest correlation of luminosity and metallicity. These statistical parallax results are summarized in table 2.1.

2.5 Main sequence fitting

2.5.1 The observer's route

The goal of main sequence fitting is to determine the distance to a star cluster by fitting the observed color–magnitude diagram for the cluster main sequence to a main sequence calibrated in absolute units (such as M_v, $(B - V)_0$). To do this most directly, one would identify unevolved nearby main sequence stars with accurate parallax determinations and with the same metal abundance as the cluster to be fitted. The nearby main sequence stars would be used to define the M_v–$(B-V)_0$ relation for a main sequence having the cluster metallicity. By comparing this calibrated main sequence with the observed main sequence of the cluster, and correcting for interstellar reddening, the distance to the cluster would be obtained. This has been called 'the observer's route' to the determination of globular cluster distances (Buonanno et al. 1989).

Recently, photometry of globular cluster stars with CCD (charge-coupled device) data has led to globular cluster color–magnitude diagrams of unprecedented detail and accuracy. Nevertheless, confidence in the usefulness of this observer's route to globular cluster distances has faded. Sandage (1970) had used the main sequence fitting technique to obtain absolute magnitudes for RR Lyrae stars in several globular clusters, but later (Sandage 1982a) rejected these results as contradicting those obtained from RR Lyrae period shifts (§2.7). Carney (1980) likewise obtained main sequence fitted distances to several globular clusters, but later (Carney et al. 1992) rejected these determinations as too uncertain to be useful.

The principal problem has been the lack of nearby subdwarf stars with both sufficiently accurate trigonometric parallaxes and the right metal abundances to allow calibrated main sequences to be established for the various globular clusters. Hanson (1979), while attempting main sequence fitting for several globular clusters, noted that only two of his calibrating subdwarfs had [Fe/H] values less than -1.2. Only the more metal-rich globular clusters have [Fe/H] values as high as -1.2, so that calibrated metal-poor main sequences have to be derived by extrapolating from the more metal-rich main sequences. Carney et al. (1992) considered only one unevolved halo subdwarf, HD 103095, to have a trigonometric parallax well enough known to fix its M_v to within ± 0.07 mag. HD 103095 has about the same metal abundance as the globular cluster M5. With this calibrating star, using the technique of main sequence fitting, Jones, Carney, and Latham (1988b) obtained $\langle M_v \rangle = 0.86 \pm 0.12$ for RR Lyrae stars in M5. This value is included in table 2.1. Similar main sequence fitting solutions for RR Lyrae stars in other globular clusters must await the determination of more accurate distances for an array of metal-poor subdwarf stars.

2.5.2   *The theoretician's route*
There is a second possible approach to main sequence fitting. One may establish the fiducial, calibrated main sequences theoretically, using model isochrones. Buonanno et al. (1990), using this approach, obtained

$$\langle M_v(\text{ZAHB}) \rangle = +0.39[\text{Fe/H}] + 1.32$$

for RR Lyraes at the ZAHB level. In deriving this result, it was assumed that the faintest RR Lyraes in a cluster were the least evolved, and hence most nearly approximated the magnitude of the ZAHB.

This relatively strong dependence of ZAHB luminosity on metallicity was also found in Sandage and Cacciari's (1990) similar analysis. Carney et al. (1992), however, found evidence for a metallicity dependent color error in the VandenBerg and Bell (1985) theoretical isochrones which were employed by Buonanno et al. Correcting for this, and using the mean magnitude rather than the ZAHB magnitude for RR Lyraes in each cluster, Carney et al. found a much shallower slope, $\Delta\langle M_v \rangle / \Delta[\text{Fe/H}] = 0.16$. King, Demarque, and Green (1988), using a different set of theoretical isochrones and a somewhat different method of analysis, also found a shallower dependency of absolute magnitude on metallicity, $\Delta\langle M_v \rangle / \Delta[\text{Fe/H}] = 0.20$. The sense of all these correlations is the same, metal-poor RR Lyrae stars are brighter than metal-rich RR Lyrae stars, but the slopes found differ by a factor of two.

2.5.3 Horizontal branch models

Related to the theoretician's approach to main sequence fitting are the horizontal branch models of Lee, Demarque, and Zinn (1990). They produced synthetic horizontal branch color–magnitude diagrams with which they attempted to explain the period shift effect (§2.7), and from which they derived an estimate of the luminosity–metallicity dependency. They obtained

$$\langle M_v \rangle = +0.17[\text{Fe/H}] + 0.82$$

for $Y = 0.23$, or

$$\langle M_v \rangle = +0.19[\text{Fe/H}] + 0.97$$

for $Y = 0.20$ (Lee 1990). These magnitudes refer to the brightnesses of typical evolved RR Lyrae stars. It is apparent that the zero-point, though not the slope, of these results is quite sensitive to the adopted helium abundance, Y. Lee (1992a) has emphasized, as well, that a linear relation between [Fe/H] and absolute magnitude may not always apply, depending upon the evolutionary state of the stars under consideration.

2.5.4 Red giant branch theory

Model evolutionary tracks for red giant stars have also been employed in an attempt to determine the slope of the absolute magnitude–metallicity relation for the RR Lyrae stars. In this approach, predictions of the luminosity of stars at the tip of the red giant branch where the helium flash occurs are compared with observations of the brightest known red giants in various globular clusters. In this way, distances to the clusters are determined, and the absolute magnitudes of the RR Lyrae stars are found. Difficulties with this method include uncertainties in unambiguously determining which star is the brightest red giant and in transforming observed colors and magnitudes to T_e and M_{bol}. By this method, Da Costa and Armandroff (1990) obtained $\Delta \langle M_v \rangle / \Delta[\text{Fe/H}] = 0.17$, though, as Carney et al. (1992) noted, it is not clear whether this refers to the mean RR Lyrae level or to the ZAHB.

Fusi Pecci et al. (1990) compared observed and theoretical locations of a bump in the differential luminosity function of globular cluster red giant branches. From this comparison, they obtained $\Delta \langle M_v(\text{ZAHB}) \rangle / \Delta[\text{Fe/H}] = 0.15\text{–}0.20$.

2.6 The Baade–Wesselink method

In 1946, Wesselink proposed a method to obtain the distance, and therefore the luminosity, of a Cepheid variable star. Expanding upon earlier suggestions of Baade (1926) and others, Wesselink proposed that, at points of equal color on the ascending and descending branches of the lightcurve, a Cepheid should have the same surface temperature and the same surface brightness. Thus, any difference in luminosity between the two phases must be attributable to differences in the radius of the star, giving the fractional change in radius $\Delta R/R$. Because the change in radius of the star over the light cycle can be measured by integrating the radial velocity curve, ΔR can also be determined directly. Combining these results gives the radius R in absolute units and allows the luminosity of the star to be calculated. Wesselink (1969) later extended the method by deriving a general relationship between surface brightness and $(B–V)$ color. Moffett (1989) has reviewed the Baade–Wesselink method as applied to both Cepheids and, in less depth, RR Lyrae variables.

Early attempts to apply the Baade–Wesselink method to the RR Lyrae stars were, however, disappointing, with inconsistent results for different phases of the light cycle. Besides limitations due to insufficiently accurate data, there appeared to be a problem with the basic assumption that equal color corresponded to equal surface brightness (Abt 1959). During the expansion phase of the pulsation cycle, the effective surface gravity of an RR Lyrae star can be considerably higher than during the contraction phase. Moreover, at least for RRab stars, shock wave phenomena during rising light can distort the color and radial velocity curves (§4.4). In the expansion phase of the pulsation cycle, the flux emitted in the visible portion of the spectrum is systematically distorted because of an apparent flux redistribution (Jones 1988). This distortion is most evident at shorter wavelengths. A result is that one cannot use $(B–V)$ photometry in applying the Baade–Wesselink approach to the RR Lyrae stars.

The history of Baade–Wesselink solutions for RR Lyrae stars is therefore one of increasing sophistication of the model stellar atmospheres used to derive temperatures and gravities, of restriction of the solutions to phase intervals avoiding the expansion phase, and of a gradual shift to longer wavelength photometry (Oke et al. 1962; Woolley and Savage 1971; Woolley and Davies 1977; McDonald 1977; Siegel 1982; Carney and Latham 1984; Burki and Meylan 1986; Cacciari et al. 1989a,b). Longmore et al. (1985) noted that for purposes of a Baade–Wesselink analysis infrared photometry of RR Lyrae stars had several advantages over visible light photometry. As noted in chapter 1, the flux in the K-band (2.2μm) is more sensitive to the radius changes of an RR Lyrae star than is the visible light flux. Second, infrared measurements are less sensitive to line blanketing. Third, infrared fluxes appear less sensitive to distortions during the expansion phase of the pulsation cycle. For these reasons, most recent Baade–Wesselink investigations for RR Lyrae stars have employed $(V–H)$, $(V–K)$, $(V–I)$, or $(V–R)$ colors, rather than $(B–V)$.

Results of recent Baade–Wesselink investigations have been summarized and critically reviewed by Jones et al. (1992), Carney et al. (1992), and Cacciari et al. (1992). Restricting ourselves to the recent work which has employed infrared photometry, we find that two basic approaches have been employed. First are variants of the surface brightness method devised by Wesselink (1969) and developed by Manduca and Bell (1981). In this approach, a photometric angular diameter, $\theta(\phi)$, is determined from the observed dereddened K-magnitudes, $(V–K)$ colors, and static model atmospheres according to the relation

$$\theta(\phi) = \mathrm{dex}\{0.2[S_0 - (m_\lambda + \mathrm{BC}) - 10 \log T_e]\}$$

where S_0 is the surface brightness constant, m_λ is the magnitude at wavelength λ, ϕ is the phase, and BC is the bolometric correction. In an iterative procedure, the θ curve is matched with the spectroscopic displacement curve, ΔR, to find the distance to the star. Phases which might cause difficulty, such as the expansion phase, are not used in the solution, though there is no unanimity as to exactly which phase intervals yield the best results. This method has been applied to nine field RR Lyrae stars by Jones et al. (1987a,b), Jones (1988), Jones et al. (1988a,b), and Jones et al. (1992). Liu and Janes (1990a) took a similar approach in their solution for 13 field RR Lyraes stars, and Cacciari et al. (1992) applied it to three more variables.

A somewhat different approach was taken by Fernley et al. (1989, 1990a,b) and Skillen et al. (1989). Cacciari et al. (1992) applied this approach as well as the surface

Table 2.2. *Mean K-band and bolometric absolute magnitudes for RR Lyrae stars*

Magnitudes	Reference
$\langle M_K \rangle = -2.33 \log P - 0.88$	Jones et al. 1992
$\langle M_K \rangle = -2.97 \log P - 1.08$	Cacciari et al. 1992
$\langle M_K \rangle = -2.38 \log P + 0.04[\text{Fe/H}] - 0.88$	Longmore 1993
$\langle M_{\text{bol}} \rangle = 0.21[\text{Fe/H}] + 1.04$	Jones et al. 1992
$\langle M_{\text{bol}} \rangle = 0.36[\text{Fe/H}] + 1.00$	Sandage 1993b

brightness technique. Initially developed by Blackwell and Shallis (1977), this so-called infrared flux method uses lightcurves at ultraviolet, visible, and infrared wavelengths to obtain the integrated flux at all phases of the pulsation cycle. Thus, unlike the surface brightness technique, the integrated flux is derived observationally, rather than from bolometric corrections based upon model stellar atmospheres. The relation

$$L(\phi) = \theta^2(\phi)\sigma T_e^4$$

together with the observed, de-reddened bolometric flux, $L(\phi)$, and a first estimate of θ allow an initial estimate of the effective temperature, T_e, at phase ϕ. The observed infrared flux is then compared to that predicted by model stellar atmospheres of that effective temperature to refine the initial estimate of θ. The solution proceeds in an iterative fashion until the final values of T_e and θ have been obtained. This approach has been used to derive absolute magnitudes for nine field RR Lyraes.

Jones et al. (1992) and Cacciari et al. (1992) used the recent Baade–Wesselink solutions to obtain M_v–[Fe/H] and M_K–[Fe/H] relations for field RR Lyrae stars. Jones et al. give an M_{bol}–[Fe/H] relation as well. These relations are given in tables 2.1 and 2.2. The relations of Jones et al. and Cacciari et al. are not, of course, independent, being based to a large measure on the same data. The M_v–[Fe/H] relation from the compilation in Cacciari et al. (1992) is shown in figure 2.1. Both Jones et al. and Cacciari et al. noted the presence of a few outlier stars in the absolute magnitude–metallicity diagram. These stars may be RR Lyraes in an unusually advanced evolutionary state compared to the other variables. Jones et al. (1992) considered that the zero-points of their relations are still uncertain by about ± 0.15 mag.

Even if a few outlying and possibly highly evolved RR Lyrae stars are excluded from the analysis, the typical RR Lyrae star used in the Baade–Wesselink solution will have evolved somewhat away from its ZAHB position. It will thus be slightly more luminous than it was on the ZAHB. Carney et al. (1992) suggested that, if the Baade–Wesselink results are to be compared with results for RR Lyraes on the ZAHB, the slope of the Baade–Wesselink $\langle M_v \rangle$–[Fe/H] relation should be increased by about 0.05.

Longmore's (1993) reanalysis of recent Baade–Wesselink results yielded the $\langle M_K \rangle$–period–[Fe/H] relation listed in table 2.2, where the period is that of the fundamental mode. The metallicity term in this relation is small and uncertain. On the other hand, the slope of the period dependency is well established and is identical to the mean slope of the $\langle m_K \rangle$–log period relation derived from observations of RR Lyraes in globular clusters.

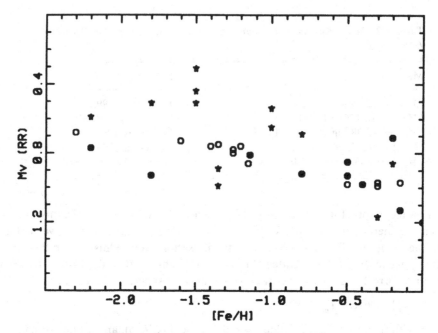

Figure 2.1 Relationship between absolute magnitude and [Fe/H] for field RR Lyrae stars analyzed by the Baade–Wesselink method, after Cacciari et al. (1992). Open circles are RR Lyraes analyzed by Liu and Janes (1990a). Closed circles are RR Lyraes analyzed by Jones and collaborators (Jones et al. 1992). 'Stars' are RR Lyraes analyzed by Fernley and collaborators (Fernley et al. 1989; 1990a,b; Skillen et al. 1989; Cacciari et al. 1992). Some points represent independent solutions for the same RR Lyrae stars.

It would, of course, be very desirable to have Baade–Wesselink analyses for RR Lyrae stars in globular clusters as well as in the field. Such analyses would help us to understand how generally applicable are the results of the field star solutions. However, even the nearest globular clusters are so distant that their RR Lyrae stars are significantly fainter than the brighter field RR Lyraes, making it more difficult to obtain the observations necessary to a Baade–Wesselink solution. As of this writing, the Baade–Wesselink solutions for globular cluster RR Lyraes (Cohen and Gordon 1987; A. Beck, in Jones et al. 1988b; Liu and Janes 1990b; Cohen 1992) are insufficient to draw sweeping conclusions, but appear in general accordance with the the analyses of field stars. It is expected that more Baade–Wesselink investigations of globular cluster RR Lyraes will.be made in the next few years.

2.7   The period shift effect
Sandage, Katem, and Sandage (1981) drew particular attention to a shift in the periods of RR Lyrae stars of equal effective temperature or equal blue amplitude in the globular clusters M3 and M15. The RR Lyrae stars in M15 ([Fe/H] = −2.1) have longer periods than their counterparts in M3 ([Fe/H] = −1.6). The period shift effect was, however, not limited to these two clusters. Sandage (1982a) found that, in general, RR Lyrae stars in metal-poor globular clusters have longer periods than those in more metal-rich globular clusters, when the RR Lyrae stars are compared at equal effective temperatures or equal amplitudes. Sandage found the magnitude of this effect to be

approximately $\Delta(\log P)/\Delta[\text{Fe/H}] = -0.12$. A period shift of the same sort has also been observed for RR Lyrae stars in the general galactic field (Sandage 1982b; Lub 1987; Sandage 1990b).

The nature of the period shift phenomenon is intimately connected with the existence of Oosterhoff groups among globular clusters and with the question of whether all globular clusters have the same age. A detailed discussion of this subject is therefore postponed until chapter 3. However, because the period shifts also appear to be related to differing RR Lyrae luminosities, they must be discussed here in brief.

As discussed in chapter 3, the period shifts can be explained if RR Lyrae stars of lower metallicity are brighter than those of higher metallicity. In their original analysis of M15 and M3 RR Lyraes, Sandage, Katem, and Sandage (1981) found that the RR Lyraes in M15 were brighter than those in M3 by 0.2 mag in m_{bol}. More recently, using masses derived from double mode RR Lyrae stars (§5.3), field RR Lyraes observed by Lub (1979), and the theoretical pulsation equations of van Albada and Baker (1971), Sandage (1990b) obtained the general relation $\Delta\langle M_v\rangle/\Delta[\text{Fe/H}] = 0.39$. Sandage's interpretation of the period shifts thus requires a strong dependency of luminosity upon metallicity. Carney et al. (1992) criticized this result on two grounds. They argued that, because the sample of RR Lyrae stars observed by Lub was specifically selected to include a wide range of metallicities and periods, it is unrepresentative of the field RR Lyraes as a whole. Second, Carney et al. argued that Sandage's method of deriving effective temperatures for the RR Lyraes was flawed. Sandage had used the mean $B-V$ colors of the RR Lyrae stars, averaged over the light cycle, as a measure of effective temperature. Carney et al, however, argued that no method of averaging the observed $(B-V)$ colors of an RR Lyrae star will give an unbiased measure of the effective temperature of the star because of excess short-wavelength emission during the expansion phase. Carney et al. obtained the shallower slope $\Delta(\log P)/\Delta[\text{Fe/H}] = -0.067$ (± 0.005) from their analysis of the period shift relation. Others have also obtained smaller slopes. Lee, Demarque, and Zinn (1990), for example, derived $\Delta(\log P)/\Delta[\text{Fe/H}] = -0.04$. These shallower slopes imply a luminosity–metallicity correlation about half as large as that suggested by Sandage, that is, about the same correlation as suggested by the Baade–Wesselink analyses.

There is general agreement, then, that the period shifts indicate that metal-poor RR Lyrae stars are more luminous than metal-rich RR Lyrae stars. However, there is a disagreement of a factor of two in the size of the effect.

Most recently, Sandage has suggested that the period shifts may have a more complex origin than a simple difference in horizontal branch luminosity. Sandage (1993a,b), in a reconsideration of the Oosterhoff phenomenon (see §3.5), found that the observed period shifts could be explained by canonical evolutionary models only if they resulted from small shifts in the temperature bounds of the instability strip, as well as luminosity. Nonetheless, his derived relations for the luminosities of RR Lyrae stars continue to show a steep dependency upon metallicity (tables 2.1, 2.2).

2.8 ω Centauri: key or red herring?

From the above, it is evident that, while there is general agreement as to the sign of the effect, there is no consensus on the size of the dependency of absolute magnitude upon metallicity. It would seem that this difficulty could be resolved were there a system so distant that all of its RR Lyraes could be regarded as essentially equally far, and which also contained RR Lyraes spanning a wide range in metallicity. The Magellanic Clouds might seem good laboratories for this sort of investigation, but RR Lyraes in each Magellanic Cloud appear to be mainly metal-poor, with only a limited range in metallicity. RR Lyraes in still more distant systems, such as M31, cannot yet be studied in the detail necessary to apply this test. There is, however, one significant test system, the globular cluster ω Centauri.

Within an individual globular cluster, all of the stars seem to have essentially the same [Fe/H]. There is one cluster, however, which departs from that rule, and that is ω Centauri. This cluster will be discussed further in chapter 3, but for the moment we may note that RR Lyrae stars within ω Centauri seem to range between [Fe/H] = −0.8 and [Fe/H] = −2.3 (Freeman and Rodgers 1975; Butler, Dickens, and Epps 1978; Gratton, Tornambe, and Ortolani 1986). The mean V magnitudes of these variables have also been determined, allowing $\langle V \rangle$ and [Fe/H] to be directly compared. The result is that there is little correlation between [Fe/H] and $\langle V \rangle$. Dickens (1989) found that most ω Centauri RR Lyraes occupy a small space in the $\langle V \rangle$, $(B–V)$ color-magnitude diagram, about 0.1 mag thick in $\langle V \rangle$, with a relatively small number of outliers. Dickens reported that four of the five most metal-rich RR Lyraes lie away from the main body of RR Lyraes in the HR diagram (three of these being fainter, and one brighter than the bulk). Otherwise, Dickens found no clear relationship between metallicity and apparent magnitude. This would appear to contradict the Baade–Wesselink and period shift results which would lead one to expect the more metal-poor ω Centauri RR Lyraes to be systematically brighter than the more metal-rich ones.

A major issue, of course, is whether ω Centauri, so unusual in its range of chemical composition, can be any sort of guide to the behavior of RR Lyrae stars in general. Is it the key to the absolute magnitude–metallicity relation, or is it a red herring – or possibly is it just one more clue to an absolute magnitude–metallicity relationship which is more complicated than was first imagined?

Gratton, Tornambe, and Ortolani (1986) argued that the particular evolutionary state of the RR Lyrae variables in ω Centauri was responsible for the lack of an overall dependency of magnitude upon metallicity. Lee (1991a) agreed with this basic idea, arguing that the ω Centauri RR Lyrae stars of −1.9 < [Fe/H] < −1.4 were highly evolved stars which had arisen from a very blue ZAHB population. Lee noted that, excluding these stars, the relationship between absolute magnitude and metallicity for the ω Centauri RR Lyraes was similar to that found by Lee, Demarque, and Zinn (1990) for the general RR Lyrae population. Among globular clusters in general, Lee argued, a much lower proportion of RR Lyraes in that metallicity range are so highly evolved. As a consequence, the relationship of luminosity and metallicity for RR Lyrae stars in ω Centauri is not identical to that for globular cluster RR Lyraes generally. There is no universal relationship between absolute magnitude and metallicity, according to Lee, because the absolute magnitude of an RR Lyrae star depends upon its evolutionary state as well as its metallicity.

2.9   Extragalactic systems

As discussed in chapter 6, RR Lyrae variables have been discovered in both Magellanic Clouds, in the dwarf spheroidal companions to the Milky Way, and in more distant Local Group galaxies. In cases where the distances to those systems can be independently determined, the absolute magnitudes of their RR Lyrae stars can be derived.

Two lines of evidence, the classical Cepheid period–luminosity relation (Feast and Walker 1987) and observations of supernova 1987A (Panagia et al. 1991; Schmidt et al. 1992), suggest that the distance modulus to the Large Magellanic Cloud is about 18.5. With this distance modulus, Walker's (1992c) observations of RR Lyrae stars in LMC star clusters give a mean RR Lyrae absolute magnitude of $\langle M_v \rangle = +0.44$ at [Fe/H] $= -1.9$. This is about 0.3 mag brighter than expected from Baade–Wesselink solutions. The LMC distance modulus may still be uncertain by perhaps ± 0.15 (see de Vaucouleurs 1993), and this may contribute to the discrepancy. Nonetheless, Walker's result suggests that at least the zero-point for the Baade–Wesselink solutions may need a small adjustment. This finding is buttressed by results from the Local Group galaxy IC 1613. If the distance modulus of IC 1613 is taken to be 24.42 ± 0.10 (Madore and Freedman 1991), which again assumes an LMC distance modulus of 18.5, the RR Lyraes stars in IC 1613 (Saha et al. 1992b) would have an absolute magnitude of about $\langle M_v \rangle = +0.46$. Adopting a slope for the absolute magnitude relation given by Baade–Wesselink solutions, and using his LMC RR Lyrae observations to set the zero-point of the relation, Walker (1992c) obtained the relation listed in table 2.1.

Simon and Clement (1993b) derived a preliminary RR Lyrae luminosity scale based upon hydrodynamic pulsation models for RRc stars. Applying their method to RRc stars in the Reticulum system, an LMC globular cluster, they obtain $\langle M_v \rangle = 0.437$, in substantial agreement with Walker's result.

Sandage (1993b), drawing upon the work of Secker (1992), noted that a possible discrepancy between the luminosity functions of globular clusters in the Galaxy and M31 may bear upon the question of RR Lyrae luminosities. The M31 clusters, with luminosities based upon the Cepheid distance scale, appear to be about 0.2 mag brighter than clusters of the Galaxy, if the luminosities of globular clusters in the Galaxy are based upon RR Lyrae absolute magnitudes derived by the Baade–Wesselink technique. This suggests that the Baade–Wesselink method may have underestimated RR Lyrae luminosities by about 0.2 mag. In a footnote to Sandage's paper, however, C. Cacciari and F. Fusi Pecci caution that the M31 cluster sample may be biased.

2.10 Astrometric distances to globular clusters

Studies of the motions of stars within globular clusters provide an additional means of determining their distances, and hence of fixing the absolute magnitudes of the RR Lyrae variables within them. In this approach, the internal velocity dispersion of stars in a globular cluster as measured from radial velocities is matched to the dispersion of observed proper motions. The accuracies of the radial velocity dispersion measurements and, especially, of the proper motion measurements are only now becoming good enough to contribute to the debate about RR Lyrae absolute magnitudes (Cudworth and Peterson 1988). Rees (1993) has derived a preliminary calibration of horizontal branch magnitudes by this technique (table 2.1). Although definitive results from this approach are not yet available, the future of this method seems promising.

2.11 Summary

In light of the above, sometimes contradictory, results, what conclusions can we draw about the absolute magnitudes of the RR Lyrae stars? It is clear that we still do not know the absolute magnitudes of the RR Lyrae stars as well as we would wish, and that to some degree the idea of *an* absolute magnitude or even the idea of a *single* absolute magnitude–metallicity relationship for all RR Lyraes is an oversimplification. It seems likely that the degree of evolution of an RR Lyrae star away from the ZAHB is one factor in fixing its absolute magnitude. Ideally, then, one should know the evolutionary history of an RR Lyrae star, including its ZAHB location, before assigning it an absolute magnitude. This, however, may often be asking for more information than is available.

The linear relation

$$\langle M_v \rangle = +0.2[\text{Fe/H}] + 0.9$$

fits many of the results of table 2.1 to within their expected uncertainties. Nonetheless, it must not be forgotten that it, and similar monotonic dependencies of luminosity upon metallicity, are simplifications which may be significantly in error for particular RR Lyrae stars.

3

RR Lyrae stars in globular clusters

Whereas the situation that held in the early decades of this century has been reversed, and the number of RR Lyrae stars now known to belong to globular clusters is dwarfed by the number of known field representatives of that species, nevertheless the globular cluster RR Lyrae stars continue to fill a vital role in the study of stellar variability. This is so not only because of convenience of observation – many cluster RR Lyraes can be recorded on a single photographic plate or CCD frame – but also because the RR Lyraes within any single globular cluster display a homogeneity not found elsewhere. The stars within any given globular cluster can be regarded as essentially coeval and, excepting the anomalous cluster ω Centauri, they appear to share a common value of [Fe/H]. These circumstances have helped to make globular clusters particularly important testing grounds for the theory of the evolution of low-mass stars. Moreover, the variable stars on the horizontal branches of globular clusters – which is to say, their RR Lyrae stars – are important objects not only to the testing of stellar evolution theory, but for the testing of stellar pulsation theory as well. For these variables, the confrontation of theory and observation has the potential to provide astronomers with information not directly accessible from studies of nonvariable cluster members.

3.1 The occurrence of RR Lyrae stars in globular clusters

The discovery of RR Lyrae stars in globular clusters, and the pioneering efforts of Solon Bailey in this endeavor, have been described in chapter 1. Although by the turn of the century, Bailey had discovered more than 500 globular cluster variables, later discoveries demonstrated that the field had by no means been exhausted. Helen Sawyer Hogg, Amelia Wehlau, Martha Hazen, L. Rosino, and numerous other investigators have more than quadrupled the number of known globular cluster variables since Bailey's day.

The growth of knowledge concerning globular cluster variables is reflected in successive editions of Helen Sawyer Hogg's *Catalog of Variable Stars in Globular Clusters*. The first edition of this catalog (Sawyer 1939) listed 1215 variable stars, of which periods were known for 656. Of those with known periods, 614 were RR Lyrae stars. In the second edition (Sawyer 1955), the number of globular cluster variables had grown to 1421. 843 of these had periods determined, and of those 779 appeared to be RR Lyrae stars. In the third catalog (Sawyer Hogg 1973), the number of globular cluster variables had risen to 2119 and the number with known periods had grown to 1313. 1202 of these 1313 are RR Lyrae stars, more than 91 percent of the total.

A more recent summary of RR Lyraes in globular clusters has been given by

Suntzeff, Kinman, and Kraft (1991). Their survey of the literature revealed that the total number of RR Lyrae stars recognized in globular clusters had risen to about 1900 by the middle of 1990. In their table 8, there are 27 globular clusters in which searches have detected no RR Lyraes and 77 globular clusters in which one or more RR Lyrae stars have been found.

By plotting the cumulative distribution of known RR Lyraes as a function of projected cluster mass, Suntzeff et al. found that fewer RR Lyraes were known near the cluster cores than were expected. This is very likely due to the greater difficulty of identifying RR Lyrae stars in the crowded central regions of the globular clusters than in the less crowded outer regions. However, assuming that searches were nearly complete in the outer regions of the clusters, Suntzeff et al. estimated that only about 6 percent of the cluster RR Lyrae stars remained to be discovered. They projected a total number of 2025 ± 30 RR Lyraes, of which perhaps 80 were really superposed field RR Lyraes, giving a total of 1945 ± 40 globular cluster RR Lyrae stars. Suntzeff et al. noted that this number is probably an underestimate, because a few globular clusters had not yet been well searched for variables and because low-amplitude variables might have been missed even in well-searched clusters. Nevertheless, they did not expect their total to be too small by more than a few percent, a result which implies that most RR Lyrae stars in globular clusters have already been discovered.

The number of RR Lyrae stars differs widely from one cluster to another. In NGC 5272 (M3) 260 RR Lyrae stars are known, whereas in a number of other well-searched clusters no RR Lyrae stars have been found. The number of RR Lyrae stars known in a globular cluster is not always a good indicator of how likely it is that an evolved star in that cluster becomes an RR Lyrae star. That is because there is a considerable range in the total number of evolved stars among the globular clusters. Following Kukarkin (1973), Suntzeff et al. normalized the observed number of RR Lyrae stars by total cluster luminosity, deriving the quantity $N(RR)$, the number of RR Lyrae stars which a globular cluster would have if it were a cluster with $M_v = -7.5$.

A plot of $N(RR)$ versus [Fe/H] is shown in figure 3.1. The cluster with the greatest value of $N(RR)$ is Pal 13, but, as only four RR Lyrae stars are known in this small cluster, the value of $N(RR)$ is subject to the uncertainties of small number statistics. Note that the value of N(RR) is small for metal-rich globular clusters, highest for moderately metal-poor globular clusters, and lower again for the most metal-deficient globular clusters. Note also that, below [Fe/H] $= -0.8$, clusters of the same [Fe/H] may have different values of $N(RR)$.

The broad trend of figure 3.1 is explicable in terms of the correlation of metallicity with horizontal branch type, the so-called first parameter governing the distribution of stars on the horizontal branch. The metal-rich globular clusters have horizontal branches which lie entirely to the red side of the instability strip in the color–magnitude diagram. $N(RR)$ is consequently near zero for these clusters. As one proceeds to more metal-poor clusters, the distribution of stars on the horizontal branch becomes bluer (figure 3.2), populating the instability strip with RR Lyrae stars and increasing the value of $N(RR)$. For the most metal-deficient globular clusters, the horizontal branch stars lie mainly to the blue side of the instability strip, and the value of $N(RR)$ decreases again.

It is clear from figure 3.1, however, that such a correlation of $N(RR)$ with metallicity cannot be the whole story. At the same value of [Fe/H] there can be a wide range in

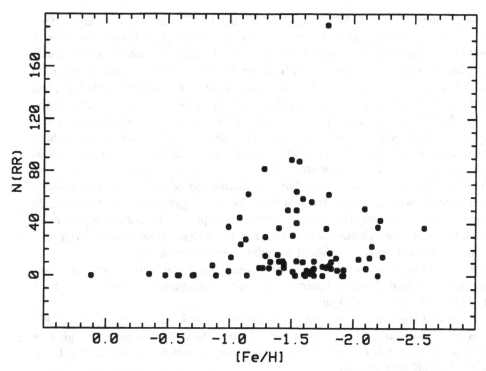

Figure 3.1 N(RR), an indicator of richness in RR Lyrae stars, is plotted against [Fe/H] for globular clusters of the Galaxy, after Suntzeff et al. (1991).

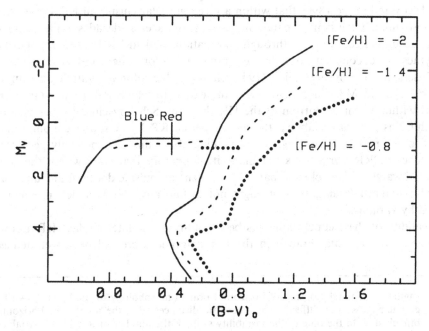

Figure 3.2 Schematic color–magnitude diagrams for globular clusters of different metallicities. Horizontal branches generally become redder with increasing metallicity, but metallicity is not the only parameter governing horizontal branch morphology. The red and blue limits to the RR Lyrae instability strip are indicated.

observed values of $N(RR)$. As Sandage and Wildey (1967) and van den Bergh (1967) were among the first to point out, horizontal branch morphology is not a function of [Fe/H] alone: clusters with the same [Fe/H] value can have horizontal branches with very different color distributions. This is the so-called second parameter problem. At least one parameter besides [Fe/H] determines horizontal branch morphology, but there has been no consensus as to what the second parameter is. Helium abundance, the abundance of CNO elements to iron, and stellar rotation have all been suggested. Zinn and co-workers (Zinn 1980; Lee, Demarque, and Zinn 1990), among others, have argued strongly that age is the most likely second parameter: those globular clusters which are older have bluer horizontal branches at a given metallicity.

Suntzeff et al. (1991) also noted that the distribution of $N(RR)$ as a function of [Fe/H] changed with galactocentric distance: 'for larger galactocentric distances, the clusters that produce more RR Lyraes are more metal-poor.' This is a consequence of a trend in horizontal branch morphology noted earlier by Zinn (1980; 1985a). At a fixed metallicity, the horizontal branches of globular clusters become redder, on average, as distance from the galactic center increases. If age is the second parameter, this would imply that inner halo globular clusters are generally older than those in the outer halo, perhaps by a few gigayears (figure 3.3).

Cacciari and Renzini (1976) produced a graphical catalog of the properties of RR Lyraes in the various globular clusters and dwarf spheroidal galaxies. Though now slightly out of date, this catalog continues to provide a useful overview of RR Lyrae properties in different systems.

3.2   The red and blue limits of the RR Lyrae instability strip

Bailey discovered in the 1890s that within a given globular cluster all RR Lyrae stars have approximately the same apparent magnitude. In the early decades of this century, evidence accumulated, primarily through observations of field RR Lyrae stars, that the RR Lyraes were confined to a limited range of color index and spectral type. Nevertheless, it was Schwarzschild's (1940) photographic color–magnitude diagram of RR Lyrae stars in M3 which made clear the extent to which RR Lyrae stars were confined within a limited portion of the HR diagram. Schwarzschild further noticed that within this zone of variability, the shorter period RR Lyrae stars were bluer than those of longer period. Surmizing that something about this zone not only permitted the existence of RR Lyrae stars but made it mandatory that stars within the zone pulsate, Schwarzschild concluded that 'a star which can pulsate does pulsate' and that 'in a color–magnitude diagram one ought not to find any non-variables in a region occupied by variables'.

The validity of these conclusions has been confirmed by later studies: RR Lyraes exist where the horizontal branch in the HR diagram is crossed by a well-defined

Figure 3.3 The relationship of horizontal branch morphology to [Fe/H] is shown for globular clusters of differing galactocentric distances. B is the number of horizontal branch stars to the blue of the instability strip, V the number of RR Lyrae variables, and R the number of horizontal branch stars to the red of the instability strip. At a given [Fe/H], cluster horizontal branches become redder with increasing galactocentric distance. The theoretical sequences show the effect of changing the cluster age by ± 2 Gyr. From Lee (1992c).

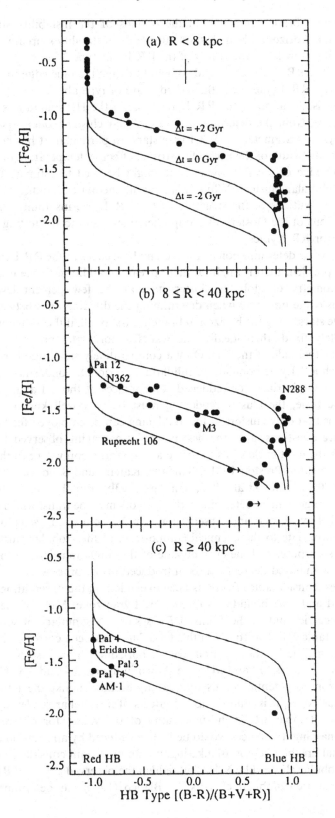

instability strip. Horizontal branch stars within the bounds of the instability strip pulsate as RR Lyrae stars. Horizontal branch stars outside the instability strip are not variable, or at least do not show large variability of the RR Lyrae type. The RRc stars are generally bluer than the RRab variables and so help to define the blue edge to the RR Lyrae instability strip. RR Lyraes near the red edge are of type ab.

Knowledge of the precise location of the RR Lyrae zone in the HR diagram is of interest for several reasons, from the determination of globular cluster reddenings to the measurement of physical parameters of RR Lyrae stars. In particular, it has been hoped that knowledge of the width of the instability strip in effective temperature or of the location of the blue edge of the RR Lyrae zone might be used to estimate the helium abundances of globular clusters. While more recent theoretical studies have dimmed these hopes, observations of the structure of the RR Lyrae instability strip remain vital to explanations of the Oosterhoff group phenomena and to the testing of stellar pulsation theory for RR Lyraes.

Observationally, the precise determination of the red and blue edges to the RR Lyrae zone is a more difficult problem than it might at first appear. In addition to the usual hazards of doing photometry of globular cluster stars at the few percent level, additional complications come into play when determining the dividing lines between variable and nonvariable stars along the horizontal branch. First of all, if these dividing lines are to be accurately fixed, then ideally one wants a horizontal branch well populated with stars on either side of the line. Only a comparatively small number of globular clusters with rich RR Lyrae populations fulfill this criterion. Second, since one generally wishes to know the intrinsic, de-reddened color indices of the red and blue edges of the RR Lyrae zone, the cluster reddening must be very well known. In practice, this often means that only clusters of low reddening can be employed for this purpose. Third, there are questions as to the best means of averaging observed RR Lyrae color indices over the light cycle so as to arrive at a quantity equivalent to that measured for nonvariable stars (Preston 1961a; Sandage, Katem, and Sandage 1981; Rood 1990; Sandage 1990a; Carney et al. 1992; Sandage 1993b). Finally, even if the colors of the red and blue edges can be determined, those colors must be transformed to effective temperatures before theory and observation can be compared. A color–temperature relation appropriate to the chemical composition of the globular cluster must be adopted for this purpose, and the current state of those relations is such that systematic errors of a few hundred degrees can be introduced in the process.

Despite these obstacles, considerable effort has been expended on the observational determination of the red and blue boundaries of the RR Lyrae zone. Most of these observations have been carried out in the Johnson *BV* system. A summary of some recent results is given in table 3.1. The colors in this table have not been corrected for differential line blanketing. The location of nonvariable horizontal branch stars and RR Lyraes in the globular clusters M3 and M15 are shown in figures 3.4 and 3.5. The width of the RR Lyrae zone is potentially a useful quantity because, unlike the actual colors of the red and blue edges, it is immune to the effects of interstellar reddening.

The dominant feature of table 3.1, is the uniformity of the values for different clusters. Some of the remaining differences would be further reduced by application of corrections for differential line blanketing. Blanketing, not an important consideration for a very metal-poor globular cluster like M15, will redden the observed colors of RR Lyrae stars in more metal-rich globular clusters. V. Blanco (1992) has determined

Table 3.1. *Red and blue edges to the RR Lyrae zone*

Cluster	$E(B-V)$	$(B-V)_{0BE}$	$(B-V)_{0RE}$	$\Delta(B-V)$	Source
N3201	0.21	0.19	0.44	0.25	Cacciari 1984
N5139 ω Cen	0.10*	0.15	0.38	0.23	Butler et al. 1978
N5272 M3	0.00	0.18	0.42	0.24	Sandage 1990a
N6723	0.00	0.17	0.43	0.26	Menzies 1974
N7078 M15	0.10	0.17	0.38	0.21	Sandage 1990a
N7078 M15	0.10	0.155	—	—	Bingham et al. 1984

*Average value; actual limits determined by applying star-by-star reddenings as in table 1 of Butler, Dickens, and Epps 1978.

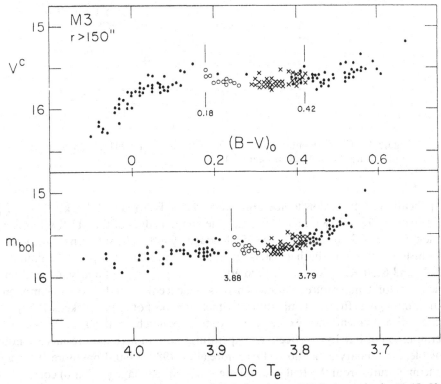

Figure 3.4 The horizontal branch of the globular cluster M3 is shown in the V, $(B-V)_0$ color–magnitude diagram and as transformed to the theoretical HR diagram, after Sandage (1990a). Filled circles represent nonvariable stars. Crosses represent variables pulsating mainly in the fundamental mode (RRab's). Open circles represent variables pulsating mainly in the first overtone mode (RRc's). Effective temperatures are based upon the Bell color–temperature calibration.

blanketing corrections, $\delta(B-V)$, for RRab stars near minimum light as a function of the ΔS metallicity parameter. The similarity of the intrinsic colors of the red and blue edges to the instability strip for different clusters has encouraged use of the observed values for these edges as a means of determining interstellar reddening for globular clusters.

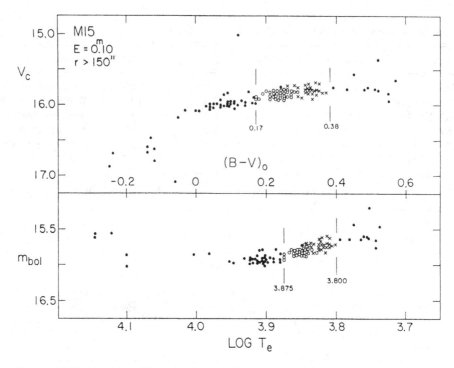

Figure 3.5 The horizontal branch of the globular cluster M15, after Sandage (1990a). Symbols are the same as in figure 3.4.

Application of the color–temperature relation of Bell (Butler, Dickens, and Epps 1978) gives log T_e = 3.88 and 3.79 for the blue and red edges of the M3 RR Lyrae zone and log T_e = 3.875 and 3.800 for M15 (Sandage 1990b). The differences between the two clusters are small; both clusters giving high and low temperature limits near 7600 K and 6250 K. However, derived effective temperatures of course depend on the choice of color–temperature relation. There is evidence that the Bell color–temperature relation may give effective temperatures which are too hot (e.g., Dickens 1989). New infrared measurements are expected to contribute much in the near future to the determination of RR Lyrae effective temperatures and, though final results are not yet available, preliminary indications (Longmore et al. 1989, 1990; Longmore 1993) agree that temperatures from the Bell relation are too high. Sandage (1993a,b) corrected the Bell color–temperature relation by making it cooler by Δlog T_e = 0.01, conforming to the temperature scale zero-point of Kurucz (1979). With this calibration, the blue edge to the RR Lyrae instability strip is located near T_e = 7400 K, while the red edge is near 6100 K.

Although table 3.1 argues for an overall uniformity to the location of the red and blue edges in different globular clusters, Sandage (1993b) has recently suggested that small, but significant, cluster to cluster differences exist. In particular, he has argued that there is a correlation of effective temperature of the instability strip with metallicity. Because this correlation is associated with his explanation for the Oosterhoff dichotomy, a fuller discussion is postponed until §3.5.

Early nonlinear pulsation models succeeded in approximately locating the instability strip in the HR diagram, but were inadequate to make precise quantitative predictions of the red and blue edges. The red edge proved particularly intractable. Both linear and nonlinear radiative models failed to return to pulsational stability with decreasing effective temperature, leading to the conclusion that convection must play an important role in quenching pulsation at low effective temperatures. RR Lyrae models by Baker and Kippenhahn (1965) and Tuggle and Iben (1973) attempted to take this into account under the assumption that time-dependent effects of convection could be ignored. However, their models still did not return to stability against pulsation near the observed red edge, and the pulsation growth rates of their models only became small at effective temperatures which would give a width to the instability strip larger than observed. The two-dimensional, time-dependent, hydrodynamic calculations of Deupree (1977a,b) yielded much more realistic results.

Deupree found that convection quenched pulsation by eliminating the mechanism by which the ionization zone stored energy during compression and released it during expansion. If one considered an 'effective opacity' which included the effects of time dependent convection, Deupree found that the 'κ mechanism' (§1.2.5) was completely negated: 'As the convective flux increases during the latter stages of contraction, the radiative opacity increases, but the effective opacity decreases with the greater efficiency of convection. The reverse is true during expansion – as the convective flux rapidly decreases, the effective opacity increases markedly.' This quenching mechanism was later confirmed by Stellingwerf (1982) and Gehmeyr (1993).

Though the inadequacies of purely radiative models for RR Lyrae stars near the red edge have been recognized for more than two decades, radiative models at first seemed more successful in defining the blue edge to the instability strip. At these higher effective temperatures, convection was less important and radiative models predicted a blue edge in approximate agreement with observation. One of the more important consequences of these models was the prediction of a strong dependence of the location of the blue edge upon helium abundance (Christy 1966; Tuggle and Iben 1972). Because the helium abundance of globular cluster stars cannot be accurately measured spectroscopically, observations of the blue edge and of the width of the RR Lyrae zone promised to provide important information on the chemical composition of globular cluster stars, and, in fact, a number of attempts were made to put the observations to this purpose (e.g. Sandage, Katem, and Sandage 1981).

Stellingwerf (1984) reinvestigated the effects of convection, however, and found that, even for the hotter RR Lyrae stars, convection played a small but significant role. In particular, Stellingwerf did not confirm the sensitive dependence of the blue edge temperature and of the width of the instability strip on helium abundance which had been obtained by purely radiative treatments. In going from $Y = 0.3$ to $Y = 0.2$, he found that the width of the RR Lyrae instability strip declined from about 1200 K to about 1000 K. In contrast, Deupree (who had maintained a radiative treatment at the blue edge) predicted a width of about 1200 K for $Y = 0.3$, but only about 150–350 K for $Y = 0.2$. These results, if correct, thus make it much more difficult to infer a helium abundance from observations of the color boundaries of the RR Lyrae zone. Stellingwerf and Bono (1993) concluded that their models constructed with a one-dimensional, nonlocal, time-dependent theory of convection seem to give a good match to observations of the RR Lyrae instability strip.

In light of pulsation theory, and of Schwarschild's original suggestion that no nonvariables ought to be found in the RR Lyrae zone, it is of interest to ask whether there are any nonvariables mixed in with the RR Lyraes in the HR diagram. In color–magnitude diagrams of globular clusters, one often finds apparently nonvariable stars within the zone occupied by the RR Lyrae stars, but most of these can be dismissed as foreground or background stars. Rarely, instances occur where there is some evidence that the nonvariable interloper may actually belong to the cluster (Geyer 1973; Smith 1985). Because of these rare cases, judgment must still be reserved concerning the universal applicability of Schwarzschild's conclusion. It may be relevant that Yao (1987) has recently reported the discovery of some very low amplitude variables in the globular cluster M4, including some near the boundaries of the RR Lyrae zone. These variables, discovered in analyses of CCD observations, would probably have escaped discovery in blink surveys of photographic plates of M4.

3.3   The dispersion in magnitude of RR Lyrae stars in individual globular clusters

The small range in apparent magnitude among the RR Lyrae stars in a given globular cluster provided a motivation for using RR Lyrae stars as standard candles. However, though the dispersion in apparent magnitude is small, it is not zero. As noted in chapter 2, evolutionary effects are expected to cause different RR Lyrae stars in the same cluster to have somewhat different absolute magnitudes. It is of interest, then, to consider whether this effect can be detected, or whether the observed dispersions in magnitude can be attributed entirely to observational error. This issue was addressed by Sandage (1990a) in an examination of RR Lyrae variables in eight relatively well-observed globular clusters.

The basic pulsation equation requires that $P\sqrt{\rho} = $ constant. Thus, excluding differences in mass, one expects that, at equal color, RR Lyraes which have longer periods should also have brighter luminosities. This is in essence the test which Sandage applied. He concluded that an intrinsic dispersion in absolute magnitude existed within each cluster. For the lowest metallicity clusters, M15 and NGC 5053, this dispersion between the brightest and faintest RR Lyraes was small, about 0.2 mag in V or m_{bol}, but the dispersion was as much as 0.6 mag for the more metal-rich cluster M4. These dispersions are large enough to be of some significance to the use of RR Lyrae stars as standard candles, particularly in circumstances where the faintest RR Lyraes in a system may be preferentially missed in discovery searches.

Sweigart (1991) compared the vertical structure of the horizontal branch, as reported by Sandage, with theoretical models of horizontal branch star evolution. He found that evolutionary tracks of stars with a helium abundance, Y, near 0.3 had more extensive blueward loops and deviated more strongly from the ZAHB than did those of stars with Y near 0.2. On the other hand, at constant Y, the evolutionary tracks of metal-rich horizontal branch stars did not deviate more strongly from the ZAHB than did those of metal-poor horizontal branch stars. Thus, canonical evolutionary models of horizontal branch stars cannot explain the trend observed by Sandage purely on the basis of the metallicity differences among the clusters.

3.4   The Oosterhoff dichotomy

That the variables in such RR Lyrae-rich clusters as M3 and ω Cen possessed different proportions of ab to c-type stars and very different period distributions was evident

Table 3.2. *Properties of Oosterhoff I and II globular clusters*

Group	$\langle P_{ab} \rangle$	$\langle P_c \rangle$	P_{tr}	$n(c)/n(ab + c)$
M3 (Oo I)	0.56	0.32	0.45	0.16
M15 (Oo II)	0.64	0.38	0.57	0.48
Oo I	0.55	0.32	0.43	0.17
Oo II	0.64	0.37	0.55	0.44

after Bailey's early work. In subsequent years, as RR Lyraes in additional globular clusters were examined, others called attention to the differing period distributions of RR Lyrae stars in different clusters (e.g. Grosse 1932; Hachenberg 1939). The name particularly associated with this subject is, however, that of P. Th. Oosterhoff. Oosterhoff (1939) drew attention to a dichotomy of properties for the RR Lyraes in a sample of five globular clusters. Three of these clusters (ω Cen, M15, and M53) had mean periods, $\langle P_{ab} \rangle$, near 0.64 days for their RRab variables and for these clusters RRc variables accounted for more than 40 percent of the total of known RR Lyraes. In contrast, two clusters (M3 and M5) had ab-type RR Lyraes with a mean period near 0.55 days and in these clusters RRc variables made up less than 20 percent of the total RR Lyrae population. Differences in the mean periods of the RRc variables in these clusters, $\langle P_c \rangle$, were also present. M3 and M5, now known as examples of Oosterhoff type I clusters, had a mean period near 0.32 days for c-type pulsators, while ω Cen, M15, and M53 (Oosterhoff type II clusters) had c-type RR Lyraes with a mean period near 0.37 days.

Though the sample of clusters originally examined by Oosterhoff was small, subsequent analyses (Oosterhoff 1944, Sawyer 1944) of RR Lyraes in 13 globular clusters soon confirmed the dichotomy, as have more recent studies (e.g. van den Bergh 1957; van Agt and Oosterhoff 1959) involving ever larger numbers of globular clusters. The very distinct period distributions of RR Lyrae stars in M3 (Oosterhoff type I) and M15 (Oosterhoff type II) are illustrated in figure 3.6. The properties of RR Lyraes in M3 and M15, and of Oosterhoff I and II clusters in general, are tabulated in table 3.2. In this table, P_{tr} indicates the transition period between RRab and RRc variables, denoted as the shortest period for an RRab pulsator in the cluster.

Arp (1955) pointed out that Oosterhoff type II clusters had steeper red giant branches in the m_v, m_{pg} color–magnitude diagram than did Oosterhoff type I clusters. He also drew attention to evidence that spectra of Oosterhoff type II clusters exhibited weaker metal lines than spectra of Oosterhoff type I clusters. This latter point was soon confirmed by Kinman (1959a). As reliable metal abundances became available for globular clusters, it became clear that these correlations reflected the circumstance that Oosterhoff type II clusters were more metal poor than those of Oosterhoff type I.

It also gradually became apparent that to regard the Oosterhoff groups as a strict dichotomy was an oversimplification. There exists a continuum of properties within each Oosterhoff group (Wehlau and Demers 1977; Sandage 1982a; Sandage 1993a). This is illustrated in figure 3.7 where, following Sandage, $\langle P_{ab} \rangle$ is plotted against [Fe/H] for globular clusters containing a significant number of RR Lyrae stars. The separation into the two Oosterhoff groups is evident, but so is a small range of $\langle P_{ab} \rangle$ values within

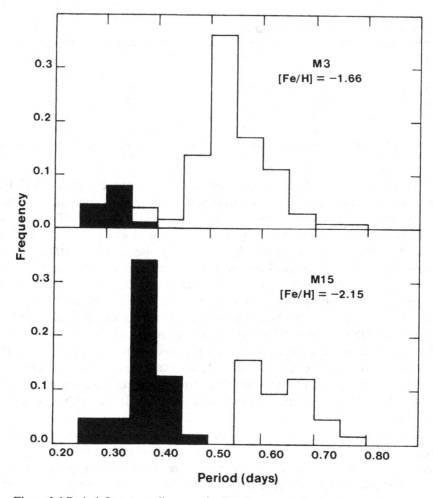

Figure 3.6 Period–frequency diagrams for RR Lyrae stars in the Oosterhoff I globular cluster M3 and the Oosterhoff II globular cluster M15. The filled area is occupied by RRc and RRd variables. The open area is occupied by RRab variables. After Kukarkin (1975). In M3 the RRab's of $P < 0.45$ day are uncertain.

each group. A similar effect is seen in the $\langle P_c \rangle$ – [Fe/H] plot (figure 3.8). The correlations between [Fe/H] and period for individual RRab and RRc stars in globular clusters are shown in figures 3.9 and 3.10, respectively.

Though there is considerable scatter in the relations, $\langle P_{ab} \rangle$ and $\langle P_c \rangle$ appear to increase slightly with decreasing metallicity for the Oosterhoff type I clusters. These clusters have $-1 > $ [Fe/H] > -1.8. There is a gap between [Fe/H] $= -1.8$ and -2.0, with clusters in that metallicity range containing few RR Lyrae stars. On the metal-poor side of this gap are the Oosterhoff type II clusters, with still longer values of $\langle P_{ab} \rangle$ and $\langle P_c \rangle$. It should be noted that these correlations between period and metallicity are not restricted to RR Lyrae stars in globular clusters. Similar correlations have been discovered to hold for RR Lyrae stars in the galactic field (§4.1).

It has sometimes been suggested that it would be useful to subdivide the Oosterhoff groups into finer categories. Castellani and Quarta (1987), for example, have proposed

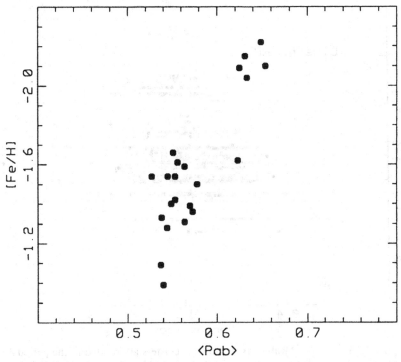

Figure 3.7 $\langle P_{ab} \rangle$ is plotted against [Fe/H] for a sample of globular clusters, after Sandage (1982a).

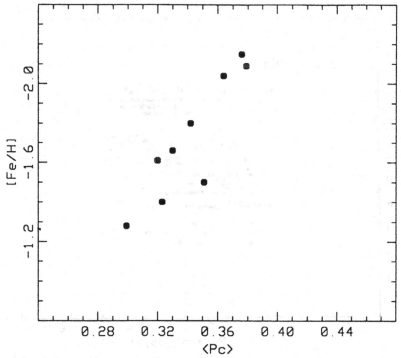

Figure 3.8 $\langle P_{c} \rangle$ is plotted against [Fe/H] for a sample of globular clusters, after Sandage (1982a).

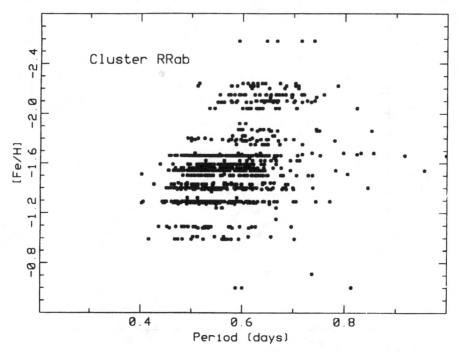

Figure 3.9 Individual RRab stars in globular clusters are plotted in the period–[Fe/H] diagram. The metallicity scale is that of Zinn (1985a), which is about 0.2 dex more metal poor than that of Butler (1975).

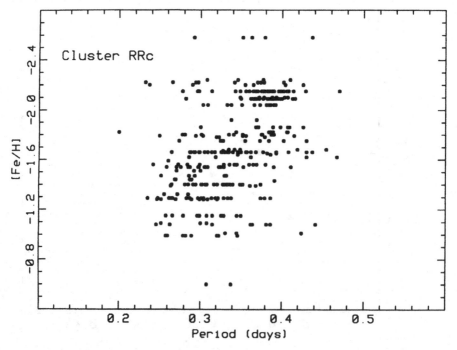

Figure 3.10 Individual RRc stars in globular clusters are plotted in the period–[Fe/H] diagram. The metallicity scale is again that of Zinn (1985a).

that the Oosterhoff type I category be subdivided into two smaller groups. Oosterhoff Ia clusters, such as M3 or NGC 7006, have [Fe/H] values near -1.6 while Oosterhoff Ib clusters, such as M5 and M62, are more metal-rich, at [Fe/H] $= -1.4$. Castellani and Quarta define the 'left ab-boundary' as the period denoted by the sharp decrease in the distribution of RRab periods at the lower edge of their distribution. They find that this quantity, which they believe to be less sensitive to cluster richness than the shortest RRab period, is different for Oosterhoff Ia, Oosterhoff Ib, and Oosterhoff type II clusters. The utility of these subdivisions is still debated, however, and none has yet come into general use.

3.5   The period shift effect and the nature of the Oosterhoff groups

Though the existence of an Oosterhoff dichotomy has been known since 1939, an adequate explanation for the phenomenon was long in coming, and in fact is still a topic of controversy. As noted by Sandage (1993a), there are three basic possibilities which could explain the mean differences in period between the Oosterhoff groups. First, if the mean effective temperatures and masses of the RR Lyrae stars in the two Oosterhoff groups are the same, then the requirement that $P\sqrt{\rho} = $ constant would imply that the RR Lyraes in the Oosterhoff type II clusters are less dense, and thus about 0.2 mag brighter, than RR Lyraes in the Oosterhoff type I clusters (Sandage 1958).

Second, there is the possibility that the RR Lyrae stars in the different Oosterhoff groups do not have the same ensemble temperatures. Schwarzschild (1958) noted that, if the entire instability gap were shifted to cooler temperatures in Oosterhoff II clusters compared to Oosterhoff I clusters, then the differences in $\langle P_{ab} \rangle$ and $\langle P_c \rangle$ between the two groups might be explained. However, this explanation foundered on the lack of observational evidence for a shift of sufficient size to explain the effect.

Third, there is the possibility that mass differs significantly between the RR Lyraes of the different Oosterhoff groups.

Efforts have been made by many people to determine which of these possibilities, or which combination of possibilities, can describe all of the observed differences in the RR Lyrae populations of the Oosterhoff groups.

Christy's (1966) pioneering pulsation models for RR Lyrae stars indicated the existence of a transition line between fundamental mode and first overtone mode RR Lyraes such that the transition period was given by the relation

$$P_{tr} = 0.057(L/L_\odot)^{0.6},$$

where the transition period is that of the shortest period RRab star. The difference in transition period between the Oosterhoff I and Oosterhoff II clusters therefore implied that RR Lyrae stars in Oosterhoff II clusters were more luminous than those in Oosterhoff I clusters by a factor of 1.38. However, as Stobie (1971) noted, such a difference in luminosity by itself would not explain the large number of RRc stars in Oosterhoff type II clusters and their relative paucity in Oosterhoff type I. Stobie therefore concluded that RR Lyraes in the two Oosterhoff groups must differ in some other fundamental property as well. He suggested that the differences between the ensemble properties of RR Lyraes in ω Centauri (Oosterhoff II) and M3 (Oosterhoff I) could be accounted for if the ω Cen RR Lyraes were more helium rich by $\Delta Y = 0.2$, or, perhaps more likely, if they were more massive by $\Delta \log M = 0.10$.

Van Albada and Baker (1973) suggested a very different interpretation of the Oosterhoff dichotomy. They rejected Christy's notion of a unique relationship between transition period and luminosity. Instead, they hypothesized the existence of a hysteresis zone in the instability strip in which the mode of RR Lyrae pulsation depended upon the direction of evolution in the HR diagram (figure 3.11). RRab variables entering the hysteresis zone from the red side continue to pulsate in the fundamental mode until they reach the blue edge of the hysteresis zone. RRc variables entering the hysteresis zone from the blue side continue to pulsate in the first overtone mode until they reach the red edge of the hysteresis zone. Van Albada and Baker found that most of the differences between the Oosterhoff groups could be explained if RR Lyrae stars in Oosterhoff II clusters were evolving mainly from blue to red, while those in Oosterhoff I clusters were evolving from red to blue. Small differences in mass or luminosity could account for any residual differences in the ensemble properties of RR Lyraes in the two groups. The existence of a hysteresis zone later found theoretical support in the pulsation calculations of Stellingwerf (1975) and Stellingwerf and Bono (1993). Because of uncertainties in cluster reddenings and observational and theoretical uncertainties in determining equilibrium colors for RR Lyrae stars, the existence of such a hysteresis zone has not yet been proven or disproven by observation.

Sandage, however, in a series of classic papers showed that the Oosterhoff phenomenon could not be explained by any hypothesis which addressed only the differences in the average properties of RR Lyrae variables. The Oosterhoff groups

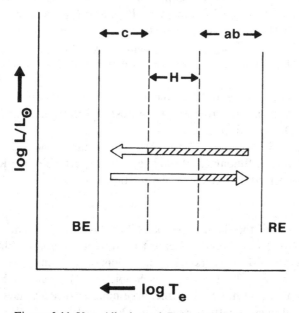

Figure 3.11 Van Albada and Baker's (1973) proposed structure for the RR Lyrae instability strip. In addition to regions of pure fundamental mode (RRab) and pure first overtone mode (RRc) pulsation, there is a hysteresis zone (labeled H) for which pulsation mode depends upon the direction of evolution. Blueward evolution, as in the left-pointing arrow, produces mainly RRab pulsators as in an Oosterhoff I cluster. Redward evolution, as in the right-pointing arrow, produces more RRc pulsators, as in Oosterhoff type II clusters.

also manifested themselves on a star-by-star basis. Sandage had initiated photographic multicolor photometry of RR Lyrae stars in the Oosterhoff I cluster M3 and the Oosterhoff II cluster M15 in part to test the hypothesis that the horizontal branches of the two clusters had different luminosities (Roberts and Sandage 1955; Sandage 1958; Sandage, Katem, and Sandage 1981). Sandage, Katem, and Sandage (1981) found a systematic shift between the M3 and M15 RR Lyraes in the amplitude–log period diagram (figure 3.12) and the rise time–log period diagram (figure 3.13). Moreover, they discovered that, as a function of color–temperature, 'the entire distributions of periods for c and ab variables again differ between M3 and M15' by that same period shift (figure 3.14). The M15 RR Lyraes had longer periods than those in M3 when compared at equal amplitudes, rise times, or color–temperatures. On the other hand, no similar shift was evident in the amplitude–color or amplitude–rise time diagrams, implying a unique relationship between these quantities.

The period of an RR Lyrae star is predicted to be a function of mass, luminosity, and effective temperature. Van Albada and Baker (1971) obtained, for example,

$$\log P = 11.497 - 0.68 \log M/M_\odot + 0.84 \log L/L_\odot - 3.48 \log T_e$$

for fundamental mode pulsators. To explain the period shifts, Sandage, Katem, and Sandage argued that RR Lyraes in M15 are more luminous than those in M3. The argument is illustrated schematically in figure 3.15, where branch A represents the M15

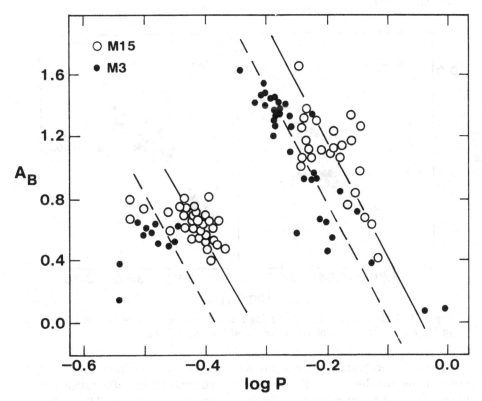

Figure 3.12 The relationship between *B* amplitude and log *P* for RR Lyrae stars in M3 and M15, after Sandage, Katem, and Sandage (1981).

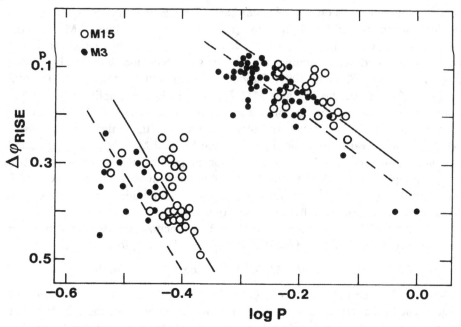

Figure 3.13 The relationship between rise time (minimum to maximum light, in phase units) and log *P* for RR Lyrae stars in M3 and M15, after Sandage, Katem, and Sandage (1981).

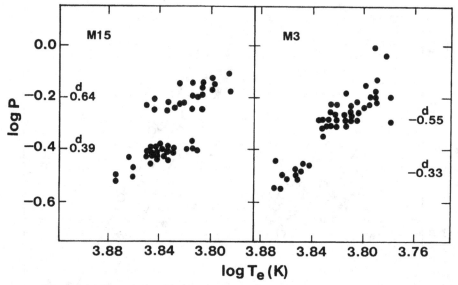

Figure 3.14 The relationship between log *P* and color–temperature for RR Lyrae stars in M3 and M15, after Sandage, Katem, and Sandage (1981).

horizontal branch and branch B the M3 horizontal branch. If the masses of the RR Lyrae stars in the two clusters are the same, we would expect a line of constant period to cut branches A and B at different effective temperatures, as shown. RR Lyraes on branch A would have longer periods than those on branch B at a given effective

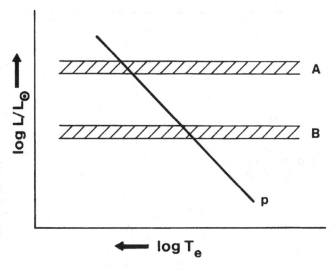

Figure 3.15 Schematic HR diagram for two horizontal branches, A and B. The line p is a line of constant period.

temperature. This is exactly the type of period shift observed by Sandage, Katem, and Sandage for M15 and M3. Sandage, Katem, and Sandage concluded that $\Delta \log L_{RR}(M15–M3) = 0.090$, when appropriate account was taken of theoretically expected small mass differences between RR Lyraes in the two clusters.

Sandage (1981, 1982a) showed that this period shift effect was not limited to M3 and M15, but that there was a general correlation of period shift with metallicity among globular cluster RR Lyraes (figure 3.16). Adopting M3 as a fiducial cluster with zero period shift, and based mainly upon comparison of blue amplitude versus log period diagrams, Sandage (1982a) obtained the relations

$$\Delta \log P = 0.116[\text{Fe/H}] + 0.173$$

and

$$\Delta m_{\text{bol}} = 0.34 \Delta[\text{Fe/H}],$$

where the sense is that the more metal-poor clusters have RR Lyraes which are brighter and of longer period at a given effective temperature. Sandage (1982b), Kemper (1982), and Lub (1987) showed that the period shift effect applied to individual field RR Lyrae stars, as well as cluster variables.

Sandage, Katem, and Sandage (1981) and Sandage (1982a) examined zero-age horizontal branch models in an attempt to explain the inferred differences in horizontal branch luminosity. They found that the differences could be explained if helium abundance and metallicity were anticorrelated so that $\Delta Y = -0.069\Delta[\text{Fe/H}]$, though they noted that the sense of the correlation seemed against intuition. Later, Sweigart, Renzini, and Tornambe (1987) verified that, with canonical ZAHB models and Los Alamos opacities, such an anticorrelation of helium and metal abundances could explain the observed period shifts, though they suspected that some as yet undiscovered effect was really the responsible agent (figure 3.17).

If there is a continuous relationship between period shift and metallicity, it may be

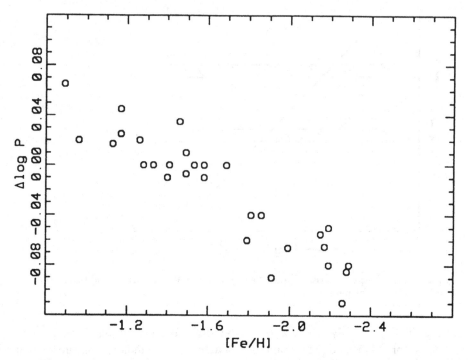

Figure 3.16 The relationship between [Fe/H] and period shift, $\Delta \log P$, is shown for RR Lyrae variables in globular clusters (after Sandage 1993b). The sense of $\Delta \log P$ is that it is negative if the RR Lyrae stars in a cluster are shifted to longer periods relative to those in M3. Note that some authors have adopted the opposite sign convention.

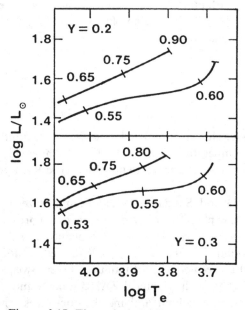

Figure 3.17 Theoretical ZAHB models showing the effects of differing helium abundance, Y. The upper sequence in each panel corresponds to $Z = 0.0001$, the lower to $Z = 0.01$. Masses of ZAHB stars are indicated in solar units. After Sweigart et al. 1987.

queried why there exists an Oosterhoff dichotomy at all, rather than an Oosterhoff continuum. This appears to be a consequence of the non-monotonic relationship between metallicity and horizontal branch morphology – i.e., the second parameter phenomenon. Renzini (1983) and Castellani (1983) pointed out that globular clusters with metallicities in the range $-2.0 < $ [Fe/H] $ < -1.7$ generally have horizontal branches which are predominantly to the blue of the instability strip, and hence have very few RR Lyrae stars. Thus, even though there might be a continuum of horizontal branch luminosities, there would be a gap in the sequence of RR Lyrae properties – with Oosterhoff type I clusters being more metal-rich than the gap and Oosterhoff type II clusters more metal-poor. Sandage (1990b; 1993a), Lee (1990), and others later endorsed this as at least a partial explanation for the Oosterhoff dichotomy, though Lee argued that the highly evolved state of RR Lyraes in Oosterhoff II clusters was also important in producing the break in $\langle P_{ab} \rangle$. A consequence of this behavior of the horizontal branch is that the relatively few RR Lyrae stars belonging to clusters with metallicities within the gap will have evolved into the instability strip from positions on the blue ZAHB. These evolved RR Lyrae stars will be brighter and have longer periods than would their ZAHB counterparts of equal effective temperature. The few RR Lyrae stars in the globular cluster M13 probably belong to this category of evolved 'gap' RR Lyraes.

Subsequent work has both extended the period shift effect and complicated its interpretation. Arguments have been made that single amplitude–effective temperature and rise time–effective temperature relations may not strictly apply (Bingham et al. 1984; Caputo 1988; Lee, Demarque, and Zinn 1990) and that there are faults in the details of Sandage's determination of the size of the period shift effect (Carney et al. 1992). Gratton et al. (1986), Caputo and De Santis (1992), Catelan (1992), and, especially, Lee, Demarque, and Zinn (1990) have argued that evolution away from the ZAHB is important to the interpretation of the period shift effect. In particular, Lee, Demarque, and Zinn (1990) have argued that most RR Lyrae stars in Oosterhoff type II clusters have evolved into the instability strip from bluer ZAHB positions. Rood (1990), however, has queried whether sufficient blue horizontal branch progenitor stars occur in M15 to feed the instability strip in this manner.

If accepted, these arguments do not eliminate the period shift effect, but they do reduce its magnitude, leading to a change in horizontal branch luminosity with [Fe/H] about half as large as Sandage originally proposed. Because this bears closely on the question of the absolute magnitude of the RR Lyrae stars, these results have already been discussed in chapter 2.

Sandage (1993a) noted that his earlier results depended upon the effective temperatures of the RR Lyrae stars, and that the determination of these effective temperatures depended upon the adopted interstellar reddenings for the various globular clusters. Taking note of criticisms that relatively small (± 0.03 mag) differences in the adopted values of $E(B-V)$ could significantly affect his results, Sandage (1993a,b) reconsidered the Oosterhoff phenomenon on the basis of observations which he believed to be less sensitive to reddening errors.

Sandage re-evaluated the slope between [Fe/H] and the ensemble averages, $\langle P_{ab} \rangle$, for field and cluster variables. He also investigated the relationship between metallicity and the periods of the shortest period field and cluster RRab stars, those which define the blue fundamental edge to the instability strip. For both cases, he found a steep relationship:

$$\Delta(\log P_{ab})/\Delta[\text{Fe/H}] = -0.12.$$

He further found that these relations could be explained by canonical horizontal branch models (a) if there was a small decrease in temperature at every place in the instability strip by $\Delta \log T_e = 0.012$ for every one dex decrease in [Fe/H], and (b) if luminosity increased with decreasing metallicity.

Sandage, Katem, and Sandage's (1981) explanation of the Oosterhoff phenomenon and period shifts relied mainly upon an increase in RR Lyrae luminosity with decreasing metallicity. Sandage's (1993b) explanation invokes changes of both luminosity and effective temperature. The resultant absolute magnitude–metallicity relation (table 2.2) remains steep. The idea that differences in average temperature may play a role in explaining the Oosterhoff phenomenon gains some support from Fourier decomposition analyses of cluster RRc stars (§3.6).

As I write, detailed pulsation models for RR Lyrae stars are rendered somewhat uncertain as the evaluation of the effects of revised opacity relations continues (see §5.3.2). Nevertheless, Simon's (1992) comparison of stellar evolution theory versus stellar pulsation theory for RR Lyrae stars in M15 may shed some light on the debate over the period shift phenomenon. Simon compared masses and effective temperatures for M15 RR Lyraes as determined by the evolutionary tracks of Lee and Demarque (1990) and the pulsation equations of Cox (1987) and Simon (1990). He found the agreement to be unsatisfactory in several respects. No single set of models could account for all RR Lyrae periods and luminosities while also predicting a ratio of blue horizontal branch stars to RR Lyrae variables similar to that observed. Simon speculated that the Lee and Demarque evolutionary tracks might be at fault, and that the horizontal branch stars of M15 might be better described by the oxygen-enhanced evolutionary models of Dorman, Lee, and VandenBerg (1991; Dorman 1992).

3.6   The curious case of ω Centauri

ω Centauri has already been alluded to in chapter 2 in connection with the relationship between absolute magnitude and metallicity among RR Lyrae stars. Here we consider other aspects of this unique globular cluster. ω Cen is one of the most luminous and populous of the globular clusters of the Galaxy. It has many RR Lyrae variables which were first studied by Bailey (1902) and, more extensively, in a classic work by Martin (1938). The color–magnitude diagram of ω Cen by Woolley (1966) showed it to have an unusually broad giant branch. Because the color of the giant branch of a globular cluster is a function of metallicity, this suggested that there existed a large star-to-star range in [Fe/H] within ω Cen.

Subsequent studies confirmed this range in metallicity, both for red giant stars and RR Lyrae stars. Freeman and Rogers (1975), Butler, Dickens, and Epps (1978), and Gratton, Tornambe, and Ortolani (1986), using variants of Preston's (1959) ΔS method, found that RR Lyrae stars in ω Cen differed by more than a factor of ten in heavy element abundance. [Fe/H] as a function of hydrogen line spectral type for RR Lyraes in this cluster is plotted in figure 3.18. No similar spread in [Fe/H] has been observed among the RR Lyrae stars of other globular clusters (Butler 1975; Smith and Butler 1978; Smith 1984a).

Although star-to-star differences in CNO abundances have been observed for red

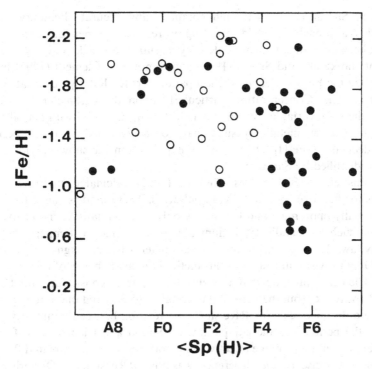

Figure 3.18 [Fe/H] versus hydrogen spectral type for RR Lyrae stars in ω Centauri, according to the observations of Butler, Dickens, and Epps (1978). Open circles are RRc stars. Filled circles are RRab stars.

giant stars in various globular clusters, ω Cen remains unique in having a large star-to-star range in heavy elements, such as calcium or iron. Because such heavy elements are not believed to be synthesized in low-mass stars, the metallicity range among the ω Cen stars is thought to be primordial, a consequence of processes active when the cluster was formed.

Dickens (1989) has presented an extensive photometric study of RR Lyrae stars in ω Cen, comparing metal abundance and photometric data for these stars. As noted in §2.8, he found no evidence for an overall dependency of luminosity upon metallicity, but there remain questions as to whether the unusual cluster ω Cen is a useful guide to the general question of the metallicity–luminosity relation. Gratton et al. (1986), Dickens (1989), and Lee (1991a) have all argued that many of the RR Lyrae stars in ω Cen have evolved from ZAHB positions on the blue horizontal branch.

3.7   Fourier decomposition

In the technique of Fourier decomposition, a Fourier series of the form

$$\text{mag} = A_0 + \Sigma \, A_j \, \cos(j\omega t + \phi_j)$$

is fitted to the observed lightcurve of a variable star. The shape of the lightcurve can then be quantified by the low order coefficients of the fit: $R_{j1} = A_j/A_1$ and $\phi_{j1} = \phi_j - j\phi_1$. The Fourier parameters have been valuable diagnostics in the comparison of observed and theoretical lightcurves for RR Lyrae stars and Cepheids (see the review in Simon 1988). Although complete and accurate lightcurves are necessary for this approach to yield meaningful results, Fourier decomposition parameters have been determined for many RR Lyrae stars in the field and in globular clusters. Clement, Jankulak, and Simon (1992) and Simon and Clement (1993) have recently investigated the Fourier decomposition parameters for RR Lyrae variables in several globular clusters. Concentrating particularly upon the parameter ϕ_{31}, they showed that, for the RRc variables, within each cluster ϕ_{31} increased with period and that clusters segregated by metallicity such that, for a given value of ϕ_{31}, period lengthened with decreasing [Fe/H]. The parameter ϕ_{31} thus seems to show a period shift effect of the sort identified by Sandage.

To interpret this effect, Simon and Clement (1993) calculated hydrodynamic pulsation models for first overtone RR Lyrae pulsators. These models suggested that ϕ_{31} depends essentially upon mass and luminosity only, and is relatively insensitive to other parameters, such as metallicity, helium abundance, effective temperature, or choice of opacity law. This may make the ϕ_{31} parameter a useful diagnostic of RRc star properties. Only provisional results of this method are available as of this writing, but Simon and Clement conclude that it yields RRc masses consistent with those derived from RRd variables (but these are still uncertain – §5.3.2) and that it may prove possible to derive a distance scale for RRc stars as a relation between luminosity and two observables, the period and the ϕ_{31} parameter. Simon and Clement also found pronounced differences in the average effective temperatures for RRc stars in different globular clusters. The sense of the difference was that RRc stars in Oosterhoff I clusters, such as M3 or M5, averaged about 200 K hotter than RRc stars in Oosterhoff II clusters such as M15 or M68.

3.8   Implications of observations of RR Lyrae stars in globular clusters

It is worth pointing out that what may appear to be relatively small differences between adopted $\langle M_v \rangle$–[Fe/H] relations for RR Lyrae variables can have very significant consequences. The difference in luminosity between the horizontal branch and the main sequence turnoff (ΔV in figure 3.19) is an important indicator of age for globular clusters. It is insensitive to the cluster reddening and is unaffected by uncertainties in the determinations of effective temperatures for the cluster stars. However, an error of only 0.1 mag in the adopted luminosity of the RR Lyrae variables can result in an error of 1.5 Gyr in the derived cluster age (Sweigart et al. 1987).

Sandage, Katem, and Sandage (1981) noted that, for most clusters, this difference was a constant $\Delta M_v = 3.5 \pm 0.1$ regardless of cluster metallicity. Sandage, Katem, and Sandage (1981) and Sandage (1982a) found that, with their steep relation between RR Lyrae·luminosity and metallicity, and the constant value of ΔV(turnoff–horizontal branch), all globular clusters had the same age to within about 10 percent. Adopting $M_v = +0.80$ for RR Lyraes in M3, Sandage (1982a) obtained an age of 17 ± 2 Gyr for the globular cluster system. This would be consistent with rapid collapse models for the formation of the galactic halo, as in Eggen, Lynden-Bell, and Sandage (1962). Sandage (1993c) re-evaluated the ages of the globular star clusters on the basis of his revised luminosity–metallicity relation (Sandage 1993a,b) and newer stellar evolutionary models incorporating enhanced oxygen abundances. He concluded that the large majority of well-observed globular clusters had ages of about 14 Gyr, noting that this result was consistent with a Hubble constant $H_0 = 45$ km/sec/Mpc and a density of the universe equal to the critical density needed to halt expansion.

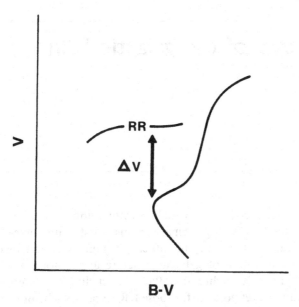

Figure 3.19 The distance ΔV between the horizontal branch and the main sequence turnoff can be a sensitive indicator of the age of a globular cluster, if the luminosity of the horizontal branch is known.

By contrast, with their more shallow slope to the luminosity–metallicity relation, Lee, Demarque, and Zinn (1990) found that the globular clusters differed significantly in age, from about 13 Gyr for relatively metal-rich clusters to about 17 Gyr for very metal-poor clusters. This would be consistent with models of galactic formation in which halo stars form over an extended time interval, as, for example, that of Searle and Zinn (1978).

Recently, Zinn (1993) and Van den Bergh (1993) have proposed that halo globular clusters can be divided into groups on the basis of horizontal branch morphology. Zinn has identified two groups of halo globular clusters: an 'Old Halo' group and a 'Younger Halo' group. The Old Halo group consists of clusters which formed during a collapse of the proto-galaxy which ended in the formation of the galactic disk. The Younger Halo group consists of clusters formed in one or more satellite systems which were later accreted by the Galaxy. At a given [Fe/H], clusters in the Old Halo group have bluer horizontal branches than clusters in the Younger Halo group. These differences in horizontal branch morphology are also reflected in differences in the RR Lyrae populations of clusters within the two groups.

Van den Bergh (1993) has proposed that the proto-Galaxy interacted with another star system that was also forming globular clusters. This 'ancestral galaxy ... formed many Oosterhoff class I clusters with M3-like color–magnitude diagrams [and] merged with the main body of the proto-Galaxy on a plunging retrograde orbit.'

Much controversy and uncertainty still exists regarding the ages of the globular clusters and the processes by which the Galaxy has evolved. The message of the RR Lyrae variables in this connection is starting to be read, but not yet with the clarity that we would wish. Nevertheless, it does seem clear that the RR Lyrae variables have an important role to play in the deciphering of the early history of the Milky Way.

4

RR Lyrae stars of the galactic field

Though much has been learned from RR Lyrae variables within globular clusters, the field RR Lyraes remain important for two principal reasons. First, with some 7000 known examples, the RR Lyraes of the field are sufficiently numerous and readily identifiable to serve as important probes of galactic structure and history. Second, because the nearest field RR Lyraes are brighter than their closest cluster counterparts, they can be studied with a detail not yet possible for those RR Lyrae stars belonging to globular clusters. Basic properties of the field RR Lyraes have been addressed in brief in chapter 1.

4.1   RR Lyrae stars in the solar neighborhood

4.1.1   Population types
Recognition of the significance of field RR Lyrae stars to some degree followed upon the realization that they could be easily identifiable standard candles. Full awareness of the importance of RR Lyrae stars as probes of the structure and history of the Galaxy had, however, to await Baade's (1944) identification of stellar populations. The occurrence of RR Lyrae stars in globular clusters and in the galactic halo at first indicated that they belonged to Baade's Population II. Other observations, however, suggested that the assumption that all RR Lyrae stars belonged to Population II was an oversimplification.

Spectroscopic studies by Munch and Terrazas (1946), Struve (1950a), and Iwanowska (1953) discovered peculiarities in spectral line strengths among the RR Lyrae variables. Struve remarked, for example, that for some field RR Lyrae stars, a spectral classification based upon the hydrogen Balmer lines yielded a later spectral type than a classification based upon the calcium K-line. Interpretation of this peculiarity was difficult because, in 1950, the general metal poverty of the Population II stars had not yet been established. Struve, led astray by some misleading information on the colors of RR Lyrae stars, tried to interpret these spectral peculiarities in terms of abnormally low hydrogen abundances, but did not find this conclusion satisfactory. Kukarkin (1949), meanwhile, had shown that short period field RRab variables ($P \approx$ 0.43 days) were more concentrated to the galactic plane than were the RR Lyraes as a whole. Moreover, such short period RRab stars did not seem to occur among the RR Lyrae stars of the globular clusters. Struve (1950b, 1951) and Kinman (1959b) found that the radial velocities of the RRab variables tended to increase with increasing period.

The significance of these disparate observations, and the place of the RR Lyrae stars among the stellar populations, were not made clear until Preston's (1959) spectroscopic survey of RR Lyrae stars brighter than $m_{pg} = 13.0$. This study, carried out with low resolution spectroscopic plates obtained with the Lick Observatory's Crossley reflector, remains a landmark in the investigation of the RR Lyrae stars, and pioneered a method of measuring RR Lyrae metal abundances which is still in use.

4.1.2   The ΔS method

Examining low-resolution blue spectra (430 Å/mm at Hγ), Preston quickly confirmed that, for spectrograms of many RR Lyrae stars, the usual MK spectral classification criteria gave discordant results when applied to different spectral lines. As Struve had noted, the spectral type as determined from the hydrogen Balmer lines was often later than the spectral type determined from the prominent CaII K-line at 3933 Å. This is to say that the K-lines were generally weaker than expected on the basis of the Balmer line spectral classification. This difference became the basis of Preston's ΔS index.

Preston classified each spectrogram so as to obtain two spectral indices for each RR Lyrae star: Sp(H), the spectral type based upon the strength of the Balmer lines of hydrogen, and Sp(K), the spectral type based upon the calcium K-line. Each spectral type was measured to a tenth of a spectral class. The index ΔS was then defined to be

$$\Delta S = 10[\text{Sp(H)} - \text{Sp}(K)].$$

Because this index varied somewhat during the light cycle of an RR Lyrae star, Preston took the standard ΔS index to be that found when the star was observed at minimum light. A spectrogram obtained at minimum light which gave classifications of Sp(H) = F5 and Sp(K) = A5 would therefore indicate a star with a ΔS value of 10, whereas had the Sp(K) value equalled F2, the ΔS value would have been 3.

Near minimum light, most RRab variables were found to have values of Sp(H) near F5 or F6 (figures 1.10 and 1.11), whereas the Sp(K) values differed widely from star-to-star, signifying a wide range in ΔS indices. Preston found that most RR Lyrae stars within 3 kpc of the Sun (which we may term the solar neighborhood) had values of ΔS between 0 and 10. Spectra of field RR Lyraes of different ΔS are shown in figure 4.1.

Preston interpreted his ΔS index as an indicator of metal abundance. This interpretation was later confirmed by analyses of high resolution spectra of RR Lyrae stars (Preston 1961b; Butler 1975) and by analyses of model stellar spectra (Manduca 1981). Calibrations of the ΔS–[Fe/H] relation appearing in the literature have differed in detail. Butler (1975) obtained:

$$[\text{Fe/H}] = -0.23 - 0.16\Delta S.$$

V. Blanco (1992), using the analyses of high resolution spectra by Butler (1975), Butler and Deming (1979), and Carney and Jones (1983), and re-evaluating ΔS measurements for the calibrating stars, obtained

$$[\text{Fe/H}] = -0.02(\pm 0.34) - 0.18(\pm 0.05)\Delta S,$$

a result similar to that of Butler, but slightly more metal-rich at the lowest values of ΔS. The Butler and Blanco scales are not, however, independent, and their good overall agreement undoubtedly reflects this fact. The calibration of Suntzeff et al. (1991) is based upon the globular cluster metallicity scale adopted by Zinn and West

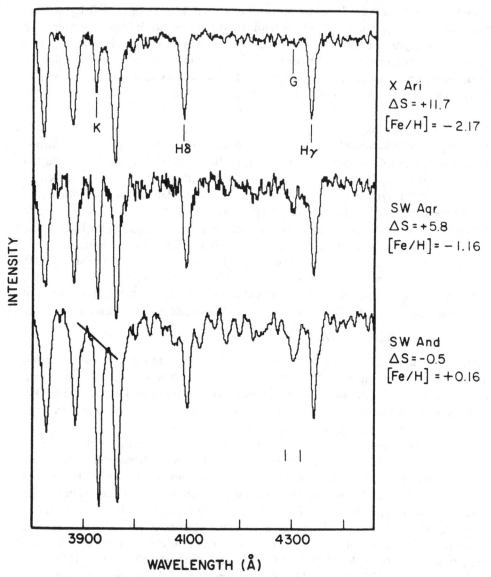

Figure 4.1 Spectra taken near minimum light of three RR Lyrae stars of different ΔS. The [Fe/H] values are from analyses of high resolution spectra. Whereas the strengths of the Balmer lines are similar in the spectra of all three stars, the metal lines (especially the calcium K-line) become progressively weaker in going from the metal-rich SW And to the very metal-poor X Ari (from Butler 1975).

(1984) and Zinn (1985a). This calibration,

$$[Fe/H] = -0.408 - 0.158\Delta S,$$

is about 0.2 dex more metal-poor than Butler's scale. These three calibrations are most discrepant for RR Lyrae stars of $0 \leq \Delta S \leq 2$.

An important aspect of these calibrations of the ΔS index is that the iron abundance [Fe/H] is inferred from the strength of the calcium K-line. To be valid, such a calibration does not require that [Fe/H] = [Ca/H]. It does, however, require that there exist a one-to-one correlation between iron and calcium abundances. In analyses of both RR Lyrae stars and nonvariable field stars in the solar vicinity, [Fe/H] and [Ca/H] appear to be well-correlated, but [Fe/H] does not equal [Ca/H]. In general, [Ca/H] appears to be slightly high compared to [Fe/H] for metal-poor stars. Manduca (1981) adopted the approximate relation [Ca/H] = 0.8[Fe/H] in his theoretical calibration of the ΔS index.

4.1.3   *Correlation of ΔS with period and motion*

Preston found that the ΔS index, and hence metallicity, correlated with both the period of an RR Lyrae star and its motion through space. For RRab variables, the average period increased with increasing ΔS (decreasing metallicity). This is shown in figure 4.2, adapted from Preston's original data. Subsequent extensions of Preston's survey have confirmed his results (figure 1.5). Preston also discovered a correlation of ΔS with the location of RRab stars in the period–amplitude diagram, an early indication of the period shift effect which Sandage (§3.5) would investigate in detail.

The short period RRab variables, which Kukarkin had noted lay close to the galactic plane, Preston found to have low values of ΔS ($\Delta S = 0–2$). These RR Lyraes, which can be termed metal-rich RR Lyraes, also have motions relative to the Sun which are small. The solar motion increases as one goes to the more metal-poor RR Lyraes, with the RR Lyraes of $\Delta S = 5–10$ having the high solar motions typical of halo stars (figure 4.3).

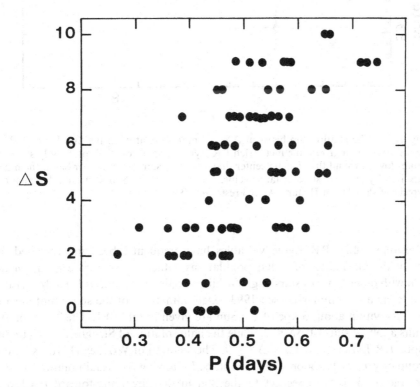

Figure 4.2 The relationship between ΔS and period for RRab stars, as determined by Preston (1959).

Table 4.1. *Velocity ellipsoids of local RR Lyrae stars*

Population	$\langle U \rangle$	$\langle V \rangle$	$\langle W \rangle$	$\sigma(U)$	$\sigma(V)$	$\sigma(W)$	Reference
Disk	+1	−42	+34	42	45	36	Taam et al. 1976
Halo	−24	−181	−6	157	104	91	Strugnell et al. 1986
$P < 0.5^{\text{d}}$	−28	−116	−2	125	125	80	Strugnell et al. 1986
$P > 0.6^{\text{d}}$	+24	−204	+8	139	102	92	Strugnell et al. 1986
Halo	—	−225	—	148	124	84	Woolley 1978

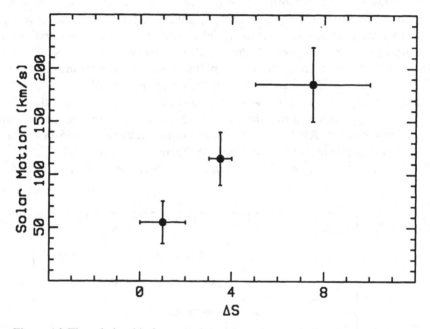

Figure 4.3 The relationship between ΔS and motion relative to the Sun for field RR Lyrae stars. As a group, the metal-rich RR Lyrae stars of low ΔS rotate only slightly more slowly around the galactic center than does the Sun. On the other hand, the high ΔS RR Lyrae stars show a large velocity relative to the Sun and have kinematics typical of Population II stars. After Preston (1959).

Since Preston's study, RR Lyrae variables have frequently been incorporated in discussions of the kinematics of stellar populations, either as a separate group or in conjunction with other types of stars (e.g. Yoshii and Saio 1979; Norris 1986; Morrison et al. 1990; Beers and Sommer-Larsen 1993). Determinations of the space motions of RR Lyrae stars within about 3 kpc of the Sun are given in table 4.1. The Taam et al. (1976) results apply to stars of $\Delta S \leq 2$, while the halo solution of Strugnell et al. (1986) is based upon RR Lyrae stars of $5 \leq \Delta S \leq 9$. The solution of Woolley (1978) is based upon a group of very metal-poor RR Lyrae stars. Following the usual notation, U, V, and W are the motions with respect to the Sun in the directions toward $l = 180°$, $b = 0°$, toward $l = 90°$, $b = 0°$, and toward $b = 90°$, respectively. The mean motions and velocity dispersions are given in units of km/sec.

4.1.4 Thick disk and halo

By the late 1950s, the concept of stellar populations had already been elaborated beyond Baade's original Populations I and II. At the Vatican conference in 1957 (O'Connell 1958), Population II had been divided into the subcategories of 'Halo Population II' and 'Intermediate Population II', while Population I had been likewise divided into 'Intermediate Population I' and 'Extreme Population I'. In between Population I and Population II was inserted a new 'disk population' of older stars with disk-like kinematics. As noted above, Preston's study demonstrated that the field RR Lyrae stars were not pure Population II, but were a mixture of stellar populations. The metal-rich field RR Lyraes belonged to an old disk population, while the more metal-poor RR Lyraes ($\Delta S = 5$–10) had motions and metallicities appropriate to the Halo Population II. This division was actually anticipated by Baade (1963), who had argued in the 1950s, even before RR Lyrae metallicities became available, that the short period RRab variables belonged to an old disk population.

In the 1980s, the idea of stellar population types was again modified with the introduction of the concept of a 'thick disk' population (see, for example, Gilmore, King, and van der Kruit 1990). The thick disk incorporates some stars that under the Vatican scheme would belong to either the disk population or intermediate Population II. It is characterized by old stars which have an exponential scale height away from the galactic plane of about 1 kpc and a rotation speed around the galactic center of about 200 km/s, lagging behind the solar rotation rate by only some 20–30 km/s. The metallicities of the thick disk stars range from approximately solar to moderately metal-poor ([Fe/H] = -0.6), but the metallicity distribution may include a tail of even more metal-poor stars (Beers and Sommer-Larsen 1993). Only one to a few percent of the stars close to the Sun belong to the thick disk; most belong to a thin disk population. The thick disk population is, as its name implies, relatively more prominent farther from the galactic plane. The globular star clusters of [Fe/H] > -0.8 have kinematics and scale height similar to that of the thick disk population (Zinn 1985a; Armandroff and Zinn 1988).

Metal-rich field RR Lyrae stars appear to belong to the thick disk population. This is perhaps clearest from the ambitious study of Layden (1993), which includes both northern and southern RR Lyraes. Layden obtained radial velocities and metallicities for a sample of 302 RRab variables within about 3 kpc of the Sun. This sample is essentially complete for RRab stars as faint as $\langle V \rangle = 12.9$ which are listed in the *General Catalogue of Variable Stars* (Kholopov et al.1985) and which have $|b| > 10°$.

For RRab's of [Fe/H] > -1, Layden's best fit exponential scale height perpendicular to the galactic plane was between 0.5 and 1 kpc, consistent with membership in the thick disk population. It is also similar to, though perhaps slightly smaller than, the scale height found for the metal-rich globular clusters (about 1.1 kpc). This relatively metal-rich group of RRab stars was also found to have a high rotation speed about the galactic center. The rotation speed for these stars, 195 ± 14 km/s, is again consistent with membership in the thick disk and is close to the rotation speed of the metal-rich globular clusters. By contrast, Layden found that field RRab's more metal-poor than [Fe/H] = -1 showed essentially no overall rotation about the galactic center. Their motions are similar to those observed for the metal-poor globular clusters classified by Zinn (1985a) as belonging to the galactic halo. Layden noted that field RR Lyraes in the metallicity range $-1.0 >$ [Fe/H] > -1.3 might be a mix of halo and thick disk populations. He further suggested that there may not be an exact match between the metallicity distributions of halo RRab's in the solar neighborhood and RRab's in the halo globular clusters.

There is some dispute over the ages of stars within the thick disk, though it is clear that it contains predominantly an old stellar population. As discussed in chapter 6, the Magellanic Clouds provide one laboratory for determining the ages of the youngest stars which can evolve to become RR Lyrae stars. In the Magellanic Clouds, the youngest cluster RR Lyraes seem to be about 12 Gyr old. If this result applies to the metal-rich RR Lyrae stars of the thick disk, then the thick disk RR Lyraes are a very old population, possibly, though not certainly, as old as the metal-rich globular clusters.

4.1.5   *RRc variables of the solar neighborhood*
Preston's original survey concentrated mainly upon RRab variables. Kemper (1982) extended the ΔS survey to a sample of RRc variables brighter than 14th magnitude. The correlation between metallicity and period for the field RRc's is shown in figure 1.6. The RRc's lack the pronounced metal-rich component of the RRab star population. This continues a trend seen among the globular cluster RR Lyrae stars: the more metal-rich Oosterhoff I globular clusters have relatively fewer RRc stars than the more metal-poor Oosterhoff II globular clusters. Kemper also determined effective temperatures for the RRc stars, based upon the strength of Balmer Hβ lines. Comparing his results with McDonald's (1977) results for RRab stars, Kemper concluded that there might exist a temperature zone in the instability strip, perhaps as wide as 600 K, in which both RRab and RRc variables exist. It is not clear whether this zone is to be identified with the hysteresis zone predicted by some models of RR Lyrae pulsation (§3.5).

With an effective temperature calibration derived from the stellar atmosphere models of Kurucz (1979), Kemper found that the hottest RRc stars had an effective temperature near 7600 K while the coolest RRab stars had an effective temperature near 6300 K. These limits are about 200 K higher than those Sandage found for globular cluster RR Lyraes (§3.2). However, because of the different temperature calibration methods employed, this difference is not large enough to signify a real difference between the bounds of the instability strip for field and cluster RR Lyraes. Kemper found that the width of the instability strip did not appear to change significantly for RR Lyrae stars of different metallicity.

4.1.6   *The 'conundrum' of the metal-rich RR Lyrae stars*
We have noted that the RR Lyrae population in the solar neighborhood includes a significant metal-rich component. Preston (1959) found that perhaps 25 percent of the RRab stars brighter than $m_{pg} = 13.0$ were relatively metal-rich ($\Delta S \leq 2$). This result has been confirmed by later studies. Of the RR Lyrae stars in the *General Catalogue* (Kholopov et al. 1985) which reach magnitude 11.0 or brighter at maximum light, Preston et al. (1991) assigned 52 stars to the halo population and 32 stars to the disk population. No decision as to population type could be reached for an additional 29 stars. All of the stars which Preston et al. considered disk stars had [Fe/H] > -1 on Butler's (1975) abundance scale ($\Delta S < 5$), and 18 of them had [Fe/H] > -0.55

($\Delta S \leq 2$). As indicated in figures 4.2 and 1.5, the metal-rich RR Lyrae stars have short periods, mostly $P < 0.45$ day.

Kraft (1972) emphasized that the existence of these metal-rich RR Lyrae stars posed a puzzle which he phrased thusly: 'Conundrum No. I: Why are the short-period Bailey a's [RRab's], which constitute about 25% of the total Bailey a population in the vicinity of the Sun, essentially unrepresented in the metal-rich globular clusters?'

The more metal-rich globular clusters, those belonging to the thick disk population, have red horizontal branches, as illustrated in figure 3.2. These horizontal branches lie entirely or almost entirely to the red side of the RR Lyrae instability strip, and, as a consequence, the metal-rich globular clusters contain few or no RR Lyrae stars. Why, then, are metal-rich RR Lyraes so common among the field RR Lyrae stars? Does this signify some fundamental difference between the field and cluster thick disk populations?

Taam, Kraft, and Suntzeff (1976) found that, in comparison with a halo red giant star, a red giant of the old disk population had only about 1/200th the probability of being detected later in its evolution as an RR Lyrae variable. This might arise in two ways. It might happen that most old disk giants do become RR Lyrae stars, but that they spend a much shorter interval of time in the RR Lyrae stage of evolution than do halo red giants. Were this the case, however, one would expect large observed period changes for metal-rich RR Lyraes. As noted in §5.1, metal-rich field RR Lyrae stars generally have low rates of period change, contradicting this prediction.

An alternative is that most disk population red giants never become RR Lyrae stars. Taam et al. constructed ZAHB models of RR Lyrae stars of solar abundance. They found that such stars would be blue enough to enter the instability strip only if they had a quite low mass, about 0.5 $M_\odot$. Because the progenitor old disk turnoff stars are expected to have much higher masses, perhaps as high as 1.0 $M_\odot$, it may be that only those rare old disk stars which undergo unusually large mass loss become metal-rich RR Lyrae stars.

The recent recognition that the metal-rich RR Lyraes belong to a thick disk population, has, however, somewhat altered this picture. The age of this population is probably greater than that of the old disk population considered by Taam et al. The masses of the progenitor turnoff stars are therefore unlikely to be greater than about 0.8 $M_\odot$, easing the mass loss problem. The size of the thick disk population is also much smaller than that of the old thin disk population. As a consequence, the probability of a thick disk red giant being detected as an RR Lyrae later in its evolution must be much higher than that given by Taam et al. for the old disk population as a whole.

4.1.7 *Completeness of surveys*

In their discussion of the field horizontal branch population of the halo, Preston et al. (1991) considered whether the discovery of nearby RR Lyrae stars has been complete. From a plot of the number of known RR Lyrae stars as a function of apparent magnitude (figure 4.4), they concluded that, for stars of magnitude $V_0 = 10.0$ or brighter there was no evidence for a magnitude dependency in the probability of discovery. Beyond 10th magnitude, incompleteness sets in, as discovery of the fainter RR Lyraes is increasingly incomplete. A plot of the number of known RR Lyraes which are brighter than $V = 11.0$ at maximum light as a function of $|b|$ indicates that

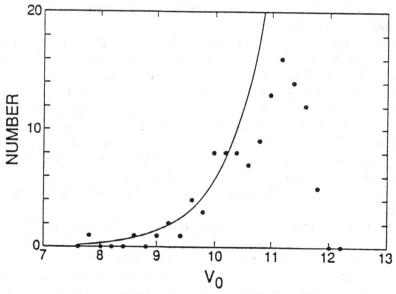

Figure 4.4 Number of field RR Lyraes as a function of apparent magnitude. The solid line shows the expected trend if the space density of RR Lyrae stars were uniform. From Preston et al. (1991).

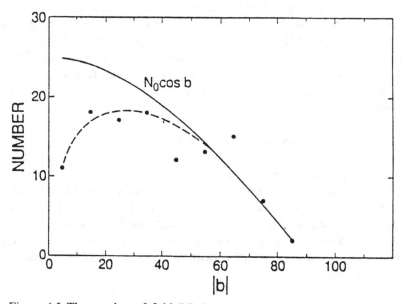

Figure 4.5 The number of field RR Lyrae stars brighter than magnitude 11.0 at maximum light (in V or B) is shown as a function of galactic latitude. The solid line shows the expected trend if the space density of RR Lyrae stars were uniform. From Preston et al. (1991).

fewer than expected RR Lyrae stars are seen at low galactic latitudes (figure 4.5). As Preston et al. pointed out, two observational considerations may explain this. Increased interstellar absorption at low galactic latitudes will dim the RR Lyrae stars. Also, discovery of RR Lyrae variables is more difficult in the crowded star fields seen at low galactic latitudes. Of course, even bright RR Lyrae stars may escape discovery if they have very low amplitudes.

4.1.8 Other metallicity indicators

Since Preston's investigation, there have been numerous studies of the metallicities of RR Lyrae stars in addition to those of Kemper and Layden. Many of these (see Smith 1984b; Clementini et al. 1991; V. Blanco 1992, and references therein) have employed the ΔS method or its variants, but other measures of RR Lyrae metallicity have also been developed. Most of these are photometric indices, among which are Sturch's (1966) broadband ultraviolet blanketing index $\delta(U–B)$, the photometric K-line index $(k–b)_2$ (Jones 1971; 1973), the Stromgren m_1 index (Epstein 1969, Epstein and Epstein 1973), and the Walraven system $\Delta[B–L]$ index (Lub 1977, 1979). The approximate relationships between ΔS and three of these photometric indices have been derived by Butler (1975) for the range $0 \leq \Delta S \leq 12$:

$$\Delta S = -84\delta(U - B) + 14$$

$$\Delta S = -44(k - b)_2 + 10$$

$$\Delta S = -99(m_1)_0 + 16.$$

The relationship between ΔS and $\Delta[B–L]$ has recently been rederived by V. Blanco (1992):

$$\Delta S = 11.9 - 194\Delta[B - L] + 815\{\Delta[B - L]\}^2.$$

4.1.9 Methods of reddening determination

Determining the color excess, usually $E(B–V)$, is a necessary prerequisite to the determination of physical parameters from multicolor photometry of RR Lyrae variables. This determination may be attempted in several ways. For RR Lyraes of high galactic latitude, a cosecant reddening law, for example, that of V. Blanco (1992), can be employed:

$$E(B–V) = 0.032 \operatorname{cosec} |b| - 0.015.$$

The cosecant law relation breaks down, however, at low galactic latitudes, and even at relatively high galactic latitudes it can occasionally be significantly in error. The alternative of determining the reddening of an RR Lyrae star from reddening determinations of neighboring nonvariable stars can be a useful approach, but also has limitations. In addition to whatever uncertainties are inherent in the reddening determination procedures for the nonvariable stars, patchiness of the interstellar reddening is a source of error, particularly at low galactic latitudes. Burstein and Heiles (1982) used observations of the galactic HI distribution to map $E(B–V)$ at latitudes more than 10 degrees from the galactic plane. These maps provide a more accurate indication of the irregularity of interstellar reddening as compared to the simple cosecant law. It should be mentioned that the Burstein and Heiles maps assume that the reddening toward both galactic poles is zero, whereas Blanco's cosecant law assumes a polar reddening of $E(B–V) = 0.02$.

Besides the general methods described above, RR Lyrae reddenings can be derived from photometry of the RR Lyrae stars themselves. Sturch (1966) investigated the determination of RRab star reddenings on the basis of broad-band UBV photometry. Following earlier suggestions, he found that near minimum light, in the phase interval $0.5 \leq \phi \leq 0.8$ after maximum light, the blanketing-corrected and reddening-corrected color, $\langle B-V \rangle_{c,0}$, was a function only of period. Thus, from the observed $(B-V)$ color, the period, and a measurement of metallicity, the color excess $E(B-V)$ could be determined.

The correlation found by Sturch between the $(U-B)_0$ and $(B-V)_0$ colors for RRab stars near minimum light is shown in figure 4.6. Sturch adopted a photometric ultraviolet blanketing index, $\delta(U-B)$, as his fundamental indicator of metallicity. This was defined by the relation

$$\delta(U-B) = (U-B) - [-0.35 + 0.72(B-V)],$$

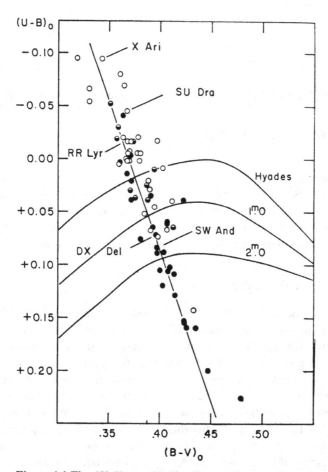

Figure 4.6 The $(U-B)_0$ vs. $(B-V)_0$ diagram for RRab stars at minimum light. Filled circles represent stars of $\Delta S = 0$–3, half-filled circles represent stars of $\Delta S = 4$–5, and open circles indicate stars of $\Delta S = 6$–10. The two color relations for Hyades main sequence stars and for stars one and two magnitudes brighter than the Hyades main sequence are also shown (from Sturch 1966).

where the $(U–B)$ and $(B–V)$ values are those observed at minimum light. A reddening slope, $E(U–B)/E(B–V) = 0.72$ is assumed.

The $\delta(U–B)$ index is a good indicator of RRab star metallicity, but is not as accurate an indicator of metallicity as is ΔS. V. Blanco (1992) reinvestigated Sturch's approach to determining reddenings, using ΔS as the indicator of metallicity and improving Sturch's blanketing corrections.

He obtained

$$E(B–V) = \langle B–V \rangle_{\phi(0.5-0.8)} + 0.0122\Delta S - 0.00045(\Delta S)^2 - 0.185P - 0.356,$$

where $\langle B–V \rangle_{\phi(0.5-0.8)}$ is the observed mean color in the 0.5–0.8 phase interval and P is the period. Blanco noted that one limitation upon the accuracy of reddenings derived in this way might be the inconstancy of RR Lyrae color curves. Very accurate lightcurves of stars not suspected of showing the Blazhko effect suggest that $\langle B–V \rangle_{\phi(0.5-0.8)}$ might change from cycle-to-cycle at the 0.01–0.02 mag level.

Blanco also re-evaluated similar methods of reddening determination in the Walraven and Stromgren photometric systems, based upon, respectively, the photometry of Lub (1977; 1979) and of Epstein (1969) and Epstein and Epstein (1973). Lub (1979), suggesting that the use of the period in deriving the reddening might be objected to, also investigated the determination of reddenings with the aid of Jones's (1973) photometry of the Balmer Hβ line. Lub (1987) noted that this index is essentially reddening free and only slightly blanketing dependent. He also noted that the Hβ method can be applied to RRc variables, unlike that of Sturch. Sperauskas (1987) has described a method for obtaining the reddening of RR Lyrae stars from observations in the Vilnaius photometric system.

4.2 Photometry, finding charts, and ephemerides

Accurate lightcurves have been determined for many of the RR Lyrae stars in the solar neighborhood and for some of the RR Lyraes in more distant locations in the Galaxy. Though these studies are too numerous to describe individually, a few comments are in order.

The *General Catalogue of Variable Stars* (Kholopov et al. 1985) and Heck's *Bibliographic Catalogue* (Heck 1988) provide basic data and references on many of the known field RR Lyrae stars. Kholopov et al. provide periods, epochs of maximum, references to finding charts, and other basic information for many stars, but some caution must be employed in using the ephemerides in the *General Catalogue*. RR Lyrae stars undergo changes in period (§5.1) and an accurate ephemeris cannot be extrapolated more than a few years into the future without a substantial possibility of error from that cause. Ephemerides for the brighter RR Lyraes are calculated at the Odessa Astronomical Observatory and published annually in the *Rocznik Astronom- iczny Obserwatorium Krakowskiego*. These ephemerides are often more up to date than those in the more infrequently revised *General Catalogue*. Tsesevich and Kazanasmas (1963) published an atlas of finding charts for field RRab stars. Many finding charts are also included in Tsesevich's (1966) book, *The RR Lyrae Stars*. Westpfahl (1992) has prepared an atlas of photographic finding charts for the brighter field RR Lyraes.

There are many diverse sources for high quality RR Lyrae lightcurves. Only a few will be mentioned here. Following Hardie's (1955) study of RR Lyrae itself, there have been numerous photoelectric studies of RR Lyrae stars in the Johnson UBV system. Sturch (1966), Fitch et al. (1966), and Bookmeyer et al. (1977) provide UBV photometry of many field RR Lyraes. Lub's (1977) photometric atlas of RR Lyrae stars observed in the Walraven system provides much useful data for bright RR Lyrae stars accessible from a southern observing site. Nikolov et al. (1984) provide a useful compendium of photoelectric data on 210 of the nearer RR Lyrae stars. Recent Baade–Wesselink studies of RR Lyrae stars (§2.4) are an excellent source of accurate lightcurves (and radial velocity curves), often including infrared, and sometimes ultraviolet, observations as well as visual photometry. Hutchinson et al. (1977), Bonnell and Bell (1985) and Bonnell et al. (1982) discuss satellite ultraviolet observations of RR Lyrae and X Arietis.

4.3   Spatial distribution and metallicities of RR Lyraes in the halo and bulge

4.3.1   *Spatial distribution*

The spatial distribution of the field RR Lyraes was studied by Shapley (1939) and Perek (1951), but the data available for those early studies were inadequate in several respects. In 1953 Walter Baade proposed a new survey of RR Lyrae variables in selected areas at galactic longitudes 0° and 180° (Blaauw 1955). The task of a new survey was taken up by Plaut (1966, 1968, 1971), who used plates taken with the 1.2 m schmidt telescope at Palomar Mountain. Plaut examined four so-called Palomar–Groningen fields, three of which were in the region of the galactic bulge at low galactic latitudes (table 4.2). The three low galactic latitude regions had been selected by Baade and Plaut for relatively low interstellar absorption.

Another extensive survey of halo RR Lyrae stars was carried out by Kinman and collaborators, who used observations obtained with the Lick Observatory astrograph. As discussed below, many of the Lick field RR Lyrae stars have been the subject of follow-up spectroscopy to determine their metallicities by the ΔS method. Studies have also been made of bulge RR Lyraes by B. Blanco (1984; 1992), and of distant halo RR Lyraes by Hawkins (1984) and Saha (1984) – see table 4.2.

These photometric studies of field RR Lyrae stars in the bulge and halo can be used to determine the distance to the galactic center and the structure of the halo. The results do, however, depend upon the completeness of the surveys, the corrections applied for interstellar absorption, and the absolute magnitude adopted for the RR Lyraes.

Kinman et al. (1965, 1966) found some evidence in their data that isodensity contours for the halo RR Lyraes were flatter in the inner halo than in its outer reaches. Oort and Plaut (1975), on the other hand, came to a different conclusion based upon analysis of the Palomar–Groningen surveys. They found that the density distribution of RR Lyraes which were located more than 1 kpc from the galactic plane was nearly spherical.

Wesselink's (1987) reinvestigation of the RR Lyrae stars in Plaut's field three has led, however, to some modification of the conclusions of Oort and Plaut. Wesselink was able to use new photographic plates of finer emulsion grain than those available to Plaut. He also was able to employ an automatic engine in measuring stellar brightnesses. An important parameter in the Oort and Plaut discussion is the completeness of Plaut's RR Lyrae surveys as a function of apparent magnitude. Plaut had known that incompleteness was a serious problem for RRc variables, and that

Table 4.2. *Some surveys of RR Lyrae variables in the halo and bulge*

Survey	l (deg)	b (deg)	B_{lim}	Field size (sq. deg)
Plaut 1966	0	+29	18	43.5
Plaut 1968	4	+12	18	43.5
Plaut 1971	0	−10	18	43.5
Wesselink 1987				
Kinman et al. 1965	11	+30	17	27.3
Kinman et al. 1966	120	+86	17	74
Hartwick et al. 1981	356	+5	19	1
Kinman et al. 1982	142	−18	17	28.4
Kinman et al. 1982	180	+26	17	28.4
Kinman et al. 1982	182	+36	17	28.4
Blanco 1984	1	−4	19	0.24
Hawkins 1984	5	−47	19	16
Saha 1984	180	+24	19.5	43.5
Saha 1984	180	+30	19.5	43.5
Saha 1984	110	−29	19.5	43.5
Stobie et al. 1986	244	−89	18	21.4
Blanco 1992	1	−5.5	19	1.25

reliable results could be expected only for variables of type RRab. However, Wesselink found that Plaut's completeness estimates were too high for the RRab stars. Re-analyzing the bulge data, with improved reddening estimates and $\langle M_B \rangle = 0.79$, Wesselink found the inner halo to be more flattened than had Oort and Plaut. In his fits to an oblate spheroid, Wesselink obtained axial ratios for isodensity contours of 0.6 ± 0.1. The density of the RR Lyraes as a function of distance from the galactic center was found to decline as a power law of index 3.0 ± 0.2, consistent with an earlier determination by Oort and Plaut. Wesselink tried fitting the data with a triaxial ellipsoid distribution, but found no reason to prefer it over the simpler oblate spheroid. Hartwick (1987) had also found it necessary to introduce a flattened component into his model of the distribution of halo RR Lyrae stars.

Preston, Shectman, and Beers (1991) reanalyzed survey data on field RR Lyrae stars. They, too, concluded that isodensity contours for RR Lyrae stars were more flattened in the inner halo than in the outer halo. To model this flattening, they adopted the relationship

$$c/a = (c/a)_0 + [1 - (c/a)_0](a/a_u) \text{ for } a < a_u$$

with $c/a = 1$ for any semi-major axis $a > a_u$. In this formula, the parameter c/a describes the axial ratio of isodensity contours perpendicular to and in the galactic plane. For an RR Lyrae absolute magnitude of $\langle M_v \rangle = +0.6$, they found satisfactory agreement with the observations for a model with a central value of $(c/a)_0 = 0.5$ and $a_u = 20$ kpc. The flattening of isodensity contours in this model is shown in figure 4.7. The RR Lyrae densities per cubic kiloparsec, ν, are represented as a power of the semi-major axis a:

$$\log \nu = 3.58 - 3.21(\pm 0.11)\log a.$$

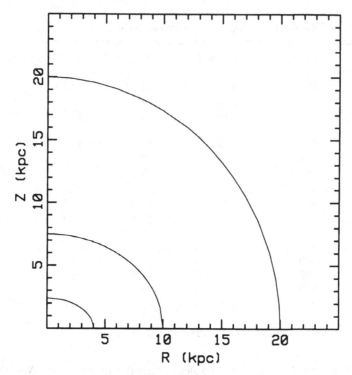

Figure 4.7 The greater flattening of isodensity contours in the inner versus the outer halo, as found by Wesselink (1987) and Preston et al. (1991). R is the distance in the galactic plane, while Z is the distance perpendicular to the plane.

Preston et al. obtained 4.8 kpc^{-3} for the local density of halo RR Lyraes, a factor of 6.5 smaller than that they obtained for field blue horizontal branch stars. They thus concluded that the mean horizontal branch described by local halo stars is quite blue. They noted, however, that the *General Catalogue of Variable Stars* (Kholopov et al. 1985) includes at least ten RR Lyraes with [Fe/H] < -1.0 within the cubic kiloparsec volume centered on the Sun. Although they commented that this might be a statistical fluctuation, they also speculated that it might indicate the existence of a metal-poor RR Lyrae population which was confined near the galactic plane, and which would thus not contribute significantly to surveys at high or intermediate galactic latitudes. Recent findings (e.g., Beers and Sommer-Larsen 1993) that the thick disk may have a significant metal-poor component may be relevant to this point.

Following Kinman et al. (1966), Suntzeff et al. (1991) also used a power law fit to the RR Lyrae density distribution, though one assuming a spherical halo:

$$\rho(R) = k(R/R_0)^n$$

with $n = -3.5$ and a local density $k = 4.5 \pm 1$ RR Lyrae per cubic kpc. Integrating this density relation for radii between 4 and 25 kpc, they estimated the total number of field RR Lyrae stars in the halo to be 85 000. Scaling the total halo mass by the RR Lyrae mass, and assuming that the field and globular cluster populations are the same, they estimated that the mass of the luminous halo is about 50 times greater than the mass of the halo globular clusters. This corresponds to a luminous halo mass of about $9 \times 10^8 \, M_\odot$ in the range 4–25 kpc from the galactic center.

4.3.2 The distance to the galactic center

RR Lyrae stars to and beyond the galactic center can be observed in a number of 'windows' located at low galactic latitude. Stars observed through these windows, of which Baade's Window ($l = -1°$, $b = -4°$) is perhaps the best known, are found to be subject to less interstellar absorption than stars in neighboring non-window fields. As Baade (1951) pointed out, the run of RR Lyrae density with distance in these window fields provides important information on the distance to the center of the system defined by the Galaxy's halo and bulge RR Lyrae stars. By inference, the center of this system is the same as the center of the Galaxy.

Adopting an RR Lyrae absolute magnitude of $\langle M_{\mathrm{pg}} \rangle = +0.7$, and using the Palomar–Groningen data described in §4.3.1, Oort and Plaut (1975) determined the distance to the galactic center to be 8.7 ± 0.6 kpc. Wesselink (1987), in his reconsideration of the bulge field data, found the somewhat smaller distance of 8.1 kpc. Walker and Terndrup (1991) analyzed photometry of Baade's Window RR Lyraes by Blanco (1984) and Walker and Mack (1986), together with their own determinations of the RR Lyrae metallicities, to find the distance to the galactic center. Adopting $\langle M_v \rangle = +0.85$ for the moderately metal-poor RR Lyraes in the Baade's Window field, they obtained a distance of 8.2 kpc. They noted, however, that their result was sensitive to the adopted corrections for interstellar absorption, and estimated ± 1 kpc to be a reasonable uncertainty. Blanco and Blanco (1985), also analyzing the Blanco (1984) and earlier data, obtained a distance to the galactic center of 8.0 ± 0.7 kpc for $\langle M_v \rangle = 0.61$. A somewhat smaller distance, 6.9 ± 0.6 kpc, was obtained assuming Sandage's (1982a) steep gradient of $\langle M_v \rangle$ with metallicity. Fernley et al. (1987) used infrared H-band (1.65 µm) observations of 70 RR Lyrae stars in Palomar–Groningen field three to obtain a distance to the galactic center of 8.0 ± 0.65 kpc. This result depends upon a period–luminosity relation in the H-band of $\langle M_{\mathrm{H}} \rangle = -0.53 - 2.0(\log P + 0.2)$. Fernley et al. noted that infrared measurements of the apparent brightnesses of bulge RR Lyraes are comparatively immune to the effects of interstellar absorption. Despite differences in detail, most recent determinations of the distance to the galactic center favor a value near 8 kpc.

4.3.3 Metallicities of RR Lyraes in the halo

The metallicities of nearly all known RR Lyrae stars brighter than 13th magnitude have been measured, sometimes by several techniques. By contrast, only a small fraction of the more distant RR Lyrae variables have had their [Fe/H] values derived by spectroscopic or photometric methods. Those surveys which have been carried out have, however, provided important clues regarding the chemical evolution of the Galaxy. Kraft and collaborators, in a companion work to the Lick photometric surveys, have obtained ΔS metallicity measures for RR Lyrae stars in the Lick fields (Butler et al. 1976; Butler et al. 1979; Butler et al. 1982a; Kinman et al. 1985; Suntzeff et al. 1991). Using these data, plus Saha and Oke's (1985) ΔS values for distant halo RR Lyraes, as well as measurements for local halo RR Lyraes, Suntzeff et al. investigated the properties of halo RR Lyrae variables.

The run of [Fe/H] versus galactocentric distance, R, is shown in figure 4.8. Suntzeff et

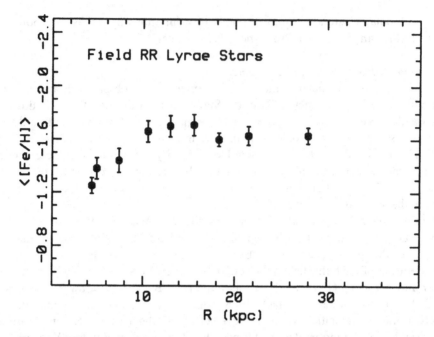

Figure 4.8 The relationship of [Fe/H] to distance from the galactic center is shown for field RR Lyrae stars. The error bars indicate the standard deviations of the means. Adapted from Suntzeff et al. (1991).

al. noted that, beyond the solar circle, the RR Lyrae population can be described as having $\langle [Fe/H] \rangle = -1.65$ with a real dispersion of about 0.30 dex, a value which does not change with increasing galactocentric distance. This result is essentially the same as found by Zinn (1985a) for globular star clusters. Inside the solar circle, on the other hand, the halo RR Lyrae stars are more metal-rich, seeming to be more metal-rich than the corresponding inner halo globular cluster population. However, Suntzeff et al. found (as had Zinn 1986) that when the observed globular cluster distribution was weighted according to the number of RR Lyrae stars in each cluster, this discrepancy vanished. This is additional evidence that the halo field RR Lyrae population is similar to the halo globular cluster RR Lyrae population.

In §3.1, it was noted that, among the globular clusters, horizontal branch morphology is a function of galactocentric distance. At a given metallicity, horizontal branches tend *on average* to be redder for clusters far from the galactic center than for those near the galactic center. This trend in horizontal branch type has an important effect upon the production of RR Lyrae stars in globular clusters. If the field RR Lyrae population is similar to the globular cluster population, as the above mentioned observations would suggest, it is plausible that the horizontal branch morphology of the field population changes with galactocentric distance in the same sense as in the globular clusters. Supporting this idea, Preston et al. (1991) concluded from observations of blue horizontal branch stars that the field horizontal branch population became redder with increasing galactocentric distance. It is also worth noting that Suntzeff et al. found that the Oosterhoff dichotomy, familiar from globular cluster studies, also was present in the field RR Lyrae stars when these were plotted in the log period–[Fe/H] diagram.

4.3.4 The metallicities of RR Lyrae stars in Baade's Window

As we saw in §4.3.2, the RR Lyrae stars in the direction of the galactic bulge have been useful in determining the distance to the center of the Galaxy, an endeavor in which the RR Lyrae stars in the direction of Baade's Window have played an important role. However, the RR Lyraes in Baade's Window are important in other ways as well, with investigations of their metallicities shedding light on the chemical evolution and history of the galactic bulge.

The metallicities of RR Lyrae stars toward Baade's Window have been probed by the ΔS method, first by Butler, Carbon, and Kraft (1976), later by Rodgers (1977) and Gratton, Tornambe, and Ortolani (1986), and most recently and extensively by Walker and Terndrup (1991). Walker and Terndrup obtained metallicities for 59 RR Lyrae stars in Baade's Window. Both the RRab and RRc stars in the sample had $\langle[Fe/H]\rangle = -1.0$ on the Zinn abundance scale. The dispersion in [Fe/H] which they derived, about 0.16 dex, was somewhat smaller than suggested by earlier studies. Walker and Terndrup noted that the RR Lyrae stars in the Baade's Window field averaged more metal-poor than the K-giant stars, many of which approach or exceed solar metallicity (Rich 1988). It is thus likely that the bulge RR Lyrae stars evolve mainly from the metal-poor end of the distribution of bulge red giants. Walker and Terndrup estimated that the probability of a K-giant turning into an RR Lyrae variable is the same for the bulge population as for the population of thick disk globular clusters.

4.3.5 Ages of RR Lyrae stars in the galactic bulge

Since it was first realized that RR Lyrae stars are evolved low-mass stars, it has been accepted that the RR Lyraes of the bulge, like those elsewhere, must be very old stars. Recently, however, Lee (1992b) has drawn upon diverse evidence to argue that the bulge RR Lyraes are the oldest RR Lyraes, and among the oldest stars of any type in the Galaxy. Lee took note of the fact that the peak of the [Fe/H] distribution for RR Lyrae stars in the bulge occurred at a higher metallicity than in the galactic halo. He further noted that the frequency of occurrence of RR Lyrae variables as a function of metallicity depends upon two things: the correlation of horizontal branch morphology with metallicity, and the metallicity distribution of the old stellar population to which the progenitors of the RR Lyrae stars belong. By constructing synthetic horizontal branches with different parameters of age and metallicity, Lee was able to conclude that the frequency distributions over [Fe/H] of RR Lyrae stars in the halo and bulge were much more sensitive to the mean correlation between horizontal branch morphology and [Fe/H] than to the abundance distribution of the underlying old population. The relatively higher metallicity of bulge RR Lyraes relative to halo RR Lyraes could be explained if the horizontal branch population of the bulge were bluer than that of the halo at a given metallicity. This is, of course, a continuation into the bulge of the second parameter phenomenon observed among the halo globular clusters (§3.1) and field horizontal branch stars (§4.3.3) of the galactic halo.

If age is the second parameter, then the bulge horizontal branch stars – and, of course, the bulge RR Lyrae stars – must be older than the oldest halo stars by about 1 Gyr, according to Lee's determination. This implies that the bulge was the first part of the Galaxy to form, and that the Galaxy was built up 'from the inside out'. Lee comments that, if this is the case, then the halo globular clusters are not the oldest objects in the Galaxy. The ages of the globular clusters as determined by stellar evolution theory are often used to set a lower limit to the age of the universe. The ages of 14–17 Gyr frequently obtained in this fashion are a matter of some concern if, as some results suggest, the value of the Hubble constant is near 90 km/s/Mpc. In that case, the inverse Hubble constant gives an expansion age to the universe of about 10 Gyr. If the bulge RR Lyrae stars are even older than the globular clusters, then the discrepancy with ages inferred from a high Hubble constant is still more extreme.

4.4   High resolution spectral studies

4.4.1   *Shock wave phenomena*

Struve (1947) and Sanford (1949) reported hydrogen line emission and line doubling at certain phases in the pulsation cycle of RR Lyrae. Similar phenomena have been observed in the spectra of other RRab variables. These spectroscopic features have generally been attributed to the existence of two shock waves during each pulsation cycle. Because of these shock waves, the radial velocity curves for RRab stars are discontinuous for the hydrogen Balmer lines and for the strong CaII absorption lines. The radial velocity curves derived from observations of weak metallic lines usually show continuity (figure 4.9). The differences in the radial velocity curves for the strong and weak lines imply that there is a velocity gradient in the atmospheres of these stars, with the higher atmospheric layers in which the strong lines are formed having the

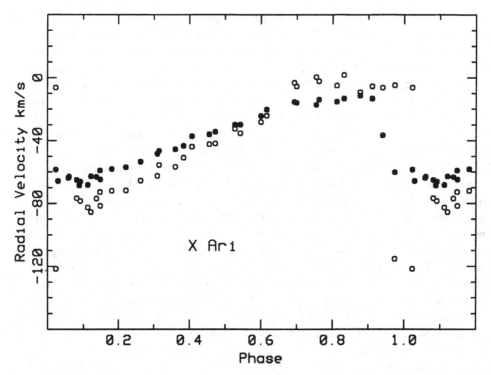

Figure 4.9 The radial velocity curve of X Ari. Filled circles indicate velocity measurements from weak metal lines. Open circles indicate velocity measurements from strong Balmer lines. Phase zero (and 1.0) corresponds to maximum light. After Oke (1966).

larger velocity amplitudes. Spectroscopic evidence for shock waves is not observed for RRc variables.

Testing models of shocks in RR Lyrae stars requires spectroscopic observations with both high spectral resolution and fine time resolution, criteria which have limited observations to a small number of relatively nearby RRab variables. This work was pioneered by Preston and collaborators (Preston and Paczynski 1964; Preston 1964; Preston et al. 1965) with photographic spectra. Recently, this type of work has been conducted by Gillet and Crowe (1988) and Gillet et al. (1989) with electronic light detectors.

The details of the line emission phenomena differ from star-to-star and, in the case of RR Lyrae itself, differ with phase in its 40.8 day secondary cycle (§5.2). In X Ari, Gillet and Crowe observed two phases of Hα emission, one centered on phase 0.76 and one on phase 0.92 (taking the phase of maximum light to be 0.00). The first emission showed a blue-shifted component, while the second, stronger, emission showed both blue and red-shifted components. The second emission episode was followed by the development of doubled Hα absorption lines near the epoch of light maximum. Line doubling did not occur after the first emission episode. Gillet et al. (1989) saw similar emission and line doubling in spectra of RX Eri, though the early emission episode was weaker than in X Ari. In spectra of RR Lyrae itself Gillet and Crowe detected weak Hα emission between phases 0.83 and 0.87, preceded by uncertain emission near phase 0.7. Line doubling after the second emission phase was observed. Preston et al. (1965) found that the hydrogen emission in the spectrum of RR Lyrae was strongest when it was at the Blazhko phase which corresponded to greatest light amplitude (§5.2). The emission was slight at Blazhko phases of low light amplitude.

Hydrodynamic models of RR Lyrae stars by Hill (1972) and Fokin (1992) agree that there are different origins for the early and main shocks. The main shock wave occurs when the infall of the photosphere halts and its outward acceleration rapidly increases. This is the shock wave responsible for the emission near phase 0.9. This shock is thus a consequence of the opacity mechanism that drives the stellar pulsation (§1.2.5). The doubled hydrogen absorption lines indicate that at certain phases this shock can divide the atmosphere into deeper outward moving layers and higher inward moving layers (Gillet and Crowe 1988). Oke et al. (1962) concluded that the main shock must be created in the atmospheric layers above those in which the weaker metallic lines are formed. The evidence for this was the continuous radial velocity curve and absence of line broadening in the metallic lines. However, Lebre (1993) has observed broadening of photospheric FeI lines in the spectrum of RR Lyrae at the times of early and main shocks. This suggests that, while the full development of the main shock may occur in the higher atmospheric layers, the shock is also detectable at the photosphere.

Gillet and Crowe considered two explanations for the early shock near phase 0.7. The hydrodynamic models suggest that the early shock results from the rapid deceleration of infalling material as the star approaches minimum radius. The 'material levitated by the previous pulsation returns to the photosphere with free fall acceleration, thus colliding with the slower inward-moving photosphere.' In this colliding' or 'infall' interpretation, the early shock should always occur before minimum radius is reached. In the second, or 'echo' model, the early shock is produced by reflection of a compression wave at the stellar core. Gillet and Crowe noted that the "echo" might be expected to produce an outward moving shock wave. Because no line doubling was observable after the early line emission episode, they concluded that the echo explanation was not likely.

Two characteristics of the lightcurves of RRab variables appear to be associated with the episodes of shock wave activity. First, at about phase 0.7, there is often a bump in the lightcurves of RRab variables. Second, there is a break in the slope on the rising branch of the lightcurve at approximately phase 0.9. This has been termed the 'hump' (Christy 1966). The 'hump' and the 'bump' in the lightcurve of RX Eri are shown in figure 4.10. For comparison, the photospheric acceleration curve and the change in radius of RX Eri are shown in figures 4.11 and 4.12.

In visible light, the hump is often just a slight hesitation in the course of light increase, but in the ultraviolet it can be seen as a distinct peak followed by a minor dip (figure 4.13). A strong ultraviolet excess is associated with the hump phases; a weak ultraviolet excess occurs at the bump phases. Gillet and Crowe (1988) noted that most RRab stars show a bump, but that it is absent in the smallest amplitude RRab variables and in RRc stars. The bump has been associated with the early shock and the hump with the main shock, though different views have been expressed as to the exact relationships of the shock waves to the variations in luminosity at these phases.

The short-lived ultraviolet excess during rising light was attributed by Hardie (1955) to hydrogen emission, by Abt (1959) to continuum radiation produced by the shock wave, and by Oke, Giver, and Searle (1962) to the increase in effective gravity during the rise to maximum. Preston (1964), though finding the issue open to debate, concluded that the Oke et al. explanation was the more likely. However, the hydrodynamic calculations by Hill and Fokin indicate that the flux contribution from

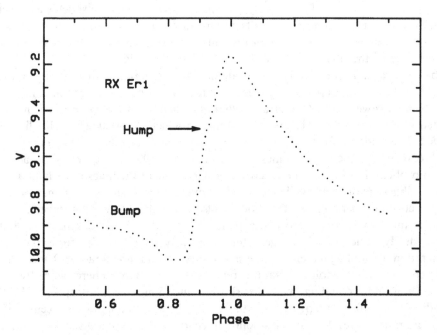

Figure 4.10 Schematic visual lightcurve of the RRab star RX Eri, after Gillet, Burki, and Crowe (1989). The locations of the 'bump' and 'hump' phases are indicated.

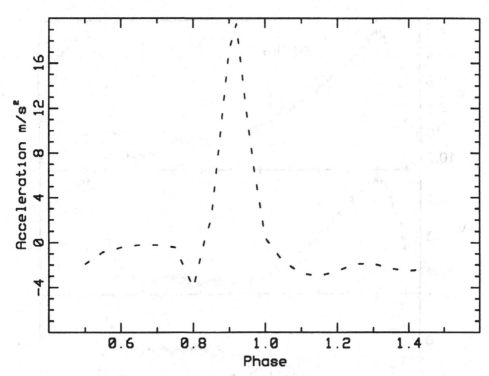

Figure 4.11 The photospheric acceleration of RX Eri is shown as a function of phase. Phase 1.0 corresponds to maximum light, as in figure 4.10. After Gillet et al. (1989).

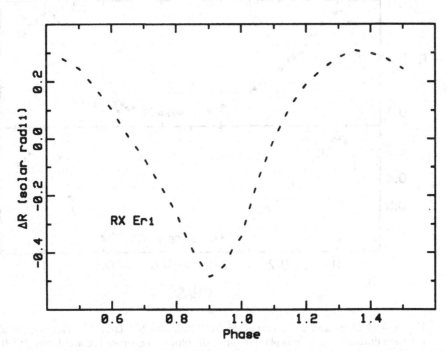

Figure 4.12 The change in radius of RX Eri is shown as a function of phase. Phase 1.0 corresponds to maximum light. After Gillet et al. (1989).

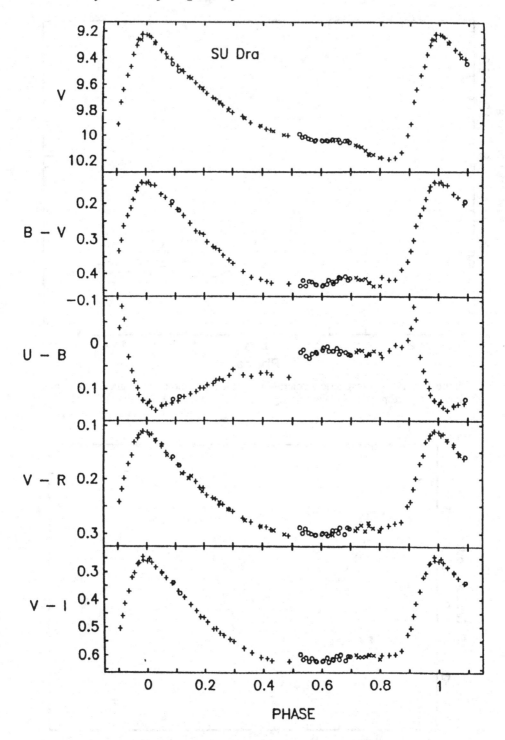

Figure 4.13 Light and color curves of the RRab star SU Dra. The excess *U*-band emission during rising light is plain in the *U–B* color curve. From Liu and Janes (1989). SU Dra is only slightly reddened by interstellar dust, with $E(B–V) = 0.02$ according to Blanco (1992).

the shock is not negligible. Gillet et al. (1989) concluded that the ultraviolet excess during rising light could be a direct consequence of the shock wave luminosity. They estimated that the effective temperature associated with the shock wave was between 20 000 and 30 000 K, corresponding to maximum emission at wavelengths of 1000 to 1500 Å. This shock emission might have only a small effect upon the visible lightcurve, but could significantly distort the lightcurve at shorter wavelengths.

On the other hand, Gillet et al. concluded that the luminosity increase during the bump phases could not be directly attributed to the weak early shock. Instead, they attributed the bump to heating of the deep atmosphere caused by the strong infall motions at these phases. In RX Eri, the actual early shock wave is produced only when the infall motion of the atmosphere is fastest, at the end of the bump phase.

Studies of the shock wave induced phenomena in RR Lyrae stars have usually been based upon observations of field RR Lyrae stars because the nearest field RR Lyraes are substantially brighter than any cluster RR Lyrae. Nevertheless, recent progress in instrumentation has made possible highly accurate photometry of the brighter cluster RR Lyrae stars. Carney et al. (1992) took advantage of high quality CCD lightcurves of RR Lyrae stars in M5 and M92 to examine aspects of the bump phenomena not easily studied with observations of field variables.

Carney et al. reconsidered the rival infall and echo explanations for the bump. The infall hypothesis, that the bump is associated with deceleration of infalling gas, is the explanation advocated by Gillet and Crowe (1988) and Gillet et al. (1989). According to the infall hypothesis, the bump must always occur before the epoch of minimum radius. Although Carney et al. did not have radial velocity data for all of the cluster variables, they noted that, based on other observations, minimum radius seems to occur at a time midway between the B light minimum and maximum. They concluded that, on the basis of times of minimum radius determined in this fashion, the bump did occur before minimum radius for all of their M5 and M92 RRab stars. However, a further test was possible for two M5 and two M92 variables for which radial velocity data were also available. If the infall model were correct, Carney et al. expected that the bump should occur later for larger expansion radii, ΔR. They found, however, no clear correlation of expansion radius and bump timing among these four stars, leading them to question the infall explanation for the bump. They also found a suggestion of a 'bump' in the lightcurve of an RRc variable in M92, which would not be predicted by the infall model. However, this bump is very small and further data are needed to confirm its reality.

In testing the rival echo hypothesis, Carney et al. assumed that the timing of the bump relative to the prior minimum radius should correlate with the equilibrium radius of the star. They noted that this hypothesis was difficult to test with field star data, because of uncertainties in relative radii. However, a test with the more homogeneous cluster star data was possible. Relative mean radii were determined for the RRab stars in each cluster from the $L \propto r^2 T_e^4$ relation. Carney et al. found a clear trend for each cluster, with the delay time for the bump being longer for larger stars. Thus, Carney et al. concluded that the echo hypothesis could not be dismissed on the basis of this test.

Oke, Giver, and Searle (1962) pointed out that the velocity gradient in the atmospheres of RR Lyrae stars poses a potential problem for attempts to determine their luminosity by the Baade–Wesselink method (§2.6). Although it is possible to exclude the hump and bump phases from the Baade–Wesselink solutions, the full radial velocity curve must be employed to find the center of mass or γ velocity of the star. Errors of only 1 or 2 km/s in this velocity can significantly affect the deduced radius and luminosity of the variable. If there is a velocity gradient in the atmosphere, then the velocity curve of any particular line may not accurately represent the motion of the photospheric layer. Jones et al. (1987a) compared the radial velocities at four phases as derived from lines of TiI and FeI with those derived from lines of TiII and FeII, which form at different depths in the lower atmosphere. They found the resultant velocities to agree to within ± 2 km/s, suggesting that any velocity gradient was small in the lower atmospheric regions in which the weak lines form. Lebre (1993), however, did detect some signs of shock waves in photospheric lines, as noted above, and it is perhaps not entirely certain that the Oke, Giver, and Searle effect has no influence on Baade–Wesselink solutions for RR Lyrae stars.

As noted above, there is no clear spectroscopic evidence for shock wave phenomena in RRc stars. However, in the lightcurves of some RRc stars there occur brief shoulders or distortions at or slightly before the phase of maximum light (as shown, for example, in the atlas of RR Lyrae lightcurves by Lub (1977)). These distortions may be associated with excess short wavelength emission. It is possible, though not certain, that these features are a result of shock wave phenomena in first overtone pulsators (Stellingwerf and Bono 1994).

4.4.2   Chemical composition

Abundances of the various chemical elements can be determined from analyses of stellar spectra of high resolution. A number of uncertainties always enter into such analyses, but, in the case of the RR Lyrae variables, two considerations have caused particular complications. First, the brightest RR Lyrae star is of the 7th magnitude and most are fainter than 9th magnitude. Observational constraints have limited abundance analyses from high resolution spectra to the brightest RR Lyrae stars. As detector efficiencies have improved in recent years, this constraint has been slightly eased. Second, RR Lyrae stars are, of course, pulsating stars. Their atmospheres are not static atmospheres, and the validity of atmospheric models calculated for static atmospheres may be questioned. Most analyses of high resolution spectra have assumed that, if the phases when the atmosphere is in most rapid motion are avoided, abundance determinations may be made as though the atmosphere were static.

Preston's (1961b) curve of growth analyses of spectra of three RR Lyraes, one of low, one of intermediate, and one of high ΔS, verified that ΔS was in fact a measure of [Fe/H]. This work was extended by Butler (1975), who obtained curves of growth abundance analyses of spectra of eleven RRab and two RRc variables. Butler used the [Fe/H] values so obtained to recalibrate the ΔS–[Fe/H] relation (§4.1.2). Butler and Deming's (1979) analyses of iron abundances in 17 RR Lyrae stars confirmed this calibration, as did the analysis of VY Ser by Carney and Jones (1983). Butler et al. (1982b) rederived iron abundances based upon the spectra of Butler and Deming, finding slightly higher but closely similar abundances.

Butler et al. (1982b; 1986) derived carbon, oxygen, and a few nitrogen abundances from high resolution spectra and uvbyβ photometry of 19 field RR Lyrae stars. These abundance determinations posed several problems. Values of [C/Fe] and [O/Fe] could only be determined from spectra taken at the hottest phases, near maximum light. Carbon abundances were determined with a local thermodynamic equilibrium (LTE) line-formation formalism, but oxygen abundances had to be determined from non-LTE analyses of the neutral oxygen triplet at 7771–7775 Å.

Carbon abundances were determined for eleven RR Lyraes of [Fe/H] > −1, with upper limits on the carbon abundance being found for eight additional RR Lyraes which were more metal poor. The [C/Fe] ratios found are similar to those seen in unevolved dwarf stars of similar metal abundance.

The results for oxygen were more puzzling. The derived [O/Fe] values are near zero for metal-poor RR Lyrae stars and show oxygen to be overabundant relative to iron for metal-rich RR Lyrae stars. These results are in contrast to those obtained for non-RR Lyrae stars in the field, where the most metal-poor giant stars seem to be relatively overabundant in oxygen. It is not yet clear, however, whether the RR Lyrae results are distorted by systematic errors in these difficult abundance determinations. Complicating the issue, measurements of the oxygen abundances among globular cluster red giant stars range from overabundant to underabundant for different stars within the same cluster (Kraft et al. 1993).

4.5 RR Lyrae stars in binary star systems

It has been estimated that at least 50 percent of Population I stars belong to binary or multiple systems (Abt 1983). Stryker et al. (1985b) found that at least 20–30 percent of the extreme subdwarf stars of Population II are within binary systems of relatively short period. There is no reason, then, not to believe that a substantial fraction of known RR Lyrae variables are members of binary or multiple star systems. Nevertheless, very few RR Lyrae stars have been established as binary stars.

This rarity of known binaries among the RR Lyrae stars undoubtedly arises in part because of the difficulty of determining whether or not an RR Lyrae star has a companion. RR Lyrae stars are old stars undergoing one of the relatively rapid stages of post-main sequence evolution. If an RR Lyrae star had a main sequence companion more massive than itself, it would be likely that the companion would have already become a white dwarf or, if it were sufficiently massive, possibly a neutron star. If the companion star were less massive than the main sequence mass of the progenitor of the RR Lyrae star, it is likely that the companion would still be on the main sequence. It is most likely, therefore, that any companion to an RR Lyrae in a binary system will be less luminous than the RR Lyrae itself. Thus, at least at visible wavelengths, binary RR Lyrae stars are unlikely to be double-line spectroscopic binaries. They may be single-line spectroscopic binaries, where the orbital motion of the RR Lyrae star is reflected in a periodic shift in its radial velocity curve. However, the orbital radial velocity shift may go unnoticed because of the large radial velocity changes due to pulsation. Only if careful allowance is made for the pulsational motion would there be a chance of detecting residual radial velocity changes due to orbital motion.

Saha and White (1990) have suggested that changes in the observed γ velocity of the RRab star TU Ursae Majoris may indicate that it is binary. They further suggested that the variations in the observed times of maximum for TU UMa might be due to light travel time effects as the star moves through its orbit. The idea that periodic variations in observed times of maxima might be attributed to orbital motion had earlier been raised in general terms by Coutts (1971). However, as discussed in §5.1, period changes and the Blazhko effect can also introduce fluctuations in times of maximum, rendering the detection of orbital motion difficult.

It has been suggested that RW Arietis is an RRc variable in an eclipsing binary system (Wisniewski 1971), but observations by Goranskij and Shugarov (1979) do not confirm the binary nature of the star.

Kinman and Carretta (1992) drew attention to two RR Lyrae stars, BB Vir and AR Her, which had bluer colors at minimum light than those expected of RRab variables. They suggested that this occurred because both stars were RRab variables with bluer companion stars. In the case of AR Her, Kinman and Caretta suggested that the companion might itself be an RR Lyrae star, but of the RRc type. Sturch (1966) had already noted the blue color of AR Her, commenting that perhaps 3 percent of RRab stars may be similar cases.

These and a few other uncertain cases are the best evidence at present for binarity among the RR Lyrae variables. It is clear, however, that further investigations are necessary before useful statistics concerning the frequency of binary star membership can be obtained for the RR Lyrae stars.

5

Period changes, the Blazhko effect, and multiple mode variables

RR Lyrae stars undergo changes which take place on a timescale longer than one pulsation period. Three types of such change are addressed in this chapter. The long term period changes of RR Lyrae variables are discussed in §5.1. Though these period changes have the potential to reveal the direction and speed of RR Lyrae evolution in the HR diagram, realizing this potential has proven difficult. Shorter term changes are considered in §5.2 and §5.3. The former discusses the Blazhko effect, a secondary cycle the existence of which has been known for eight decades, but which is still without a fully accepted explanation. The latter discusses RR Lyrae variables which pulsate simultaneously in the fundamental and first overtone radial modes. These double mode pulsators have potential for determining the masses of RR Lyrae stars, providing an important check upon masses obtained from stellar evolution theory.

5.1   Period changes

The aspect of an RR Lyrae star which can be determined with the highest accuracy is the pulsation period. With a modest number of observations, the period can be found to one part in a million or even one part in ten million. As a consequence, small changes in the pulsation period of an RR Lyrae star have the potential to reveal changes in the structure of the star long before they can be detected in any other measurable quantity. The period is related to the structure of the star by the pulsation equation $P\sqrt{\rho} = Q$. At fixed mass, the period change is therefore related to the radius change by

$$P^{-1}(dP/dt) = 1.5R^{-1}(dR/dt).$$

The potential of period change observations was realized soon after the recognition that Cepheids and RR Lyrae stars were pulsating variables. Eddington (1918) wrote: 'It would be of great interest to determine the change in period (if any) of these stars, some of which have been under observation for many years; because this would give a means of measuring a very slight change of density, and so determine the rate of stellar evolution and the length of life of a star.'

Some seven decades have passed since Eddington penned that statement and much effort has been devoted to measuring the rates of period change of RR Lyrae stars. Yet, Eddington's hope that period changes might prove the key to measuring the evolution of pulsating stars has been largely, though perhaps not completely, frustrated in the case of the RR Lyrae variables. The problem is that any evolutionary period changes in these stars tend to be masked by a period change 'noise' of irregular character.

5.1.1   *Evolutionary period changes: expectations*

Beginning in the 1960s, theorists developed reasonably specific models to describe the evolution through the HR diagram of stars on the horizontal branch (Iben and Rood 1970; Sweigart and Gross 1976). A result of these models was the prediction of the period changes expected in RR Lyrae stars because of evolution. The details of these predicted evolutionary changes depend upon a number of things. These include the masses and chemical compositions of the stars being modeled, and to some degree the details of the particular computer code employed in the calculations. Nevertheless, the two basic types of evolutionary tracks noted by Sweigart and Renzini (1979) illustrate the main expectations of evolutionary theory. The first track type, typical of metal-poor stars with initial Y near 0.3, shows slow evolution to the blue, away from the zero-age horizontal branch position, followed by more rapid evolution to the red at slightly higher luminosity. If the horizontal branch star lies within the instability strip for the duration of this evolutionary process, the resultant period changes will be similar to those shown in figure 5.1. The brief but sharp period decrease after 100 million years is associated with a rapid outgrowth of the convective core due to an instability of its helium distribution. Instabilities of this type may occur more than once toward the end of the star's horizontal branch lifetime. It is not certain whether these instabilities, sometimes called 'breathing pulses', actually occur in nature or whether they are artifacts of the methods used in calculating the internal structures of stars during this phase of their evolution (Sweigart 1991). The second type of evolutionary track is typical of metal-poor horizontal branch stars with a lower initial helium abundance, near $Y = 0.2$. Such a star evolves redward from the start, and thus is not expected to undergo the long interval of slow period decrease shown in figure 5.1. Nevertheless, the basic scenario of period change for the second type of track is similar to that of the first: an initial long interval of slow period change, followed by a more rapid increase in period as the star nears the end of its horizontal branch life. As in the first case, this interval of period increase may be interrupted by a brief time of period instability due to breathing pulses. The total horizontal branch lifetime is roughly 100 million years in both cases. Of course, stars can spend much or all of their horizontal branch lives outside of the instability strip, so that RR Lyrae star period changes may tell only a part of the evolutionary story.

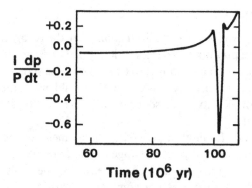

Figure 5.1 Theoretical period change per million years, $P^{-1}dP/dt$, as a function of horizontal branch lifetime for an RR Lyrae star of initial $Y = 0.3$, $Z = 0.001$, and $M = 0.7 \, M_\odot$, after Sweigart and Renzini (1979).

These evolutionary calculations lead us nonetheless to two predictions. First, most horizontal branch stars are expected to be in the relatively slow early stages of evolution away from the zero-age horizontal branch, with the consequence that, if they are RR Lyrae stars, their period changes will be small, $|dP/dt| < 0.025$ days per million years. Only in the relatively brief final stages of their horizontal branch lives, when they are moving rapidly to the red in the HR diagram, would larger period changes be expected, and those should be primarily period increases. Second, the rate of evolutionary period change alters so slowly that, over the span of a century, it is essentially constant.

5.1.2 O–C diagrams

The principle tool in the determination of period changes for RR Lyrae stars is the *O–C* diagram. The *O–C* diagram is a plot showing the observed times of maximum light (*O*) minus those calculated according to an adopted ephemeris (*C*) as a function of time. Sometimes the same information is presented on a phase diagram instead of a classical *O–C* diagram. In that case, an adopted ephemeris is used to calculate the phases of observed times of maximum light. These, rather than *O–C* values, are then plotted as a function of time. Occasionally, some fiducial point in the lightcurve other than maximum light, such as a fixed magnitude on the rising branch, is employed.

The *O–C* diagram for a star with no measurable change in period is a straight line. If the period of the star is constant, and if the correct period has been adopted, points on the *O–C* diagram will scatter about a straight horizontal line. The size of the scatter is an indication of the accuracy of the observed times of maximum or of the short term stability of the lightcurve. If the period of the star is constant, but the adopted period is too long or too short, the straight line will slope up or down (figure 5.2). If the period of the star is changing at a slow, constant rate (i.e., $P(t) = P_0 + \beta t$ where β is small), then to a good approximation the *O–C* diagram can be represented by a parabola (figure 5.3). Note that in order to interpret an *O–C* diagram correctly, it is necessary to know how many cycles have elapsed between two observed maxima. This is not always easy to determine if there have been large gaps in the observational record and if the period of the star has been changing significantly.

Random variations in the length of individual cycles about a constant mean can also in principle accumulate so as to give the impression of a spurious change in period. This is an important problem in the determination of period changes for Mira variables (Lloyd 1992). Fortunately, it is less of a problem for RR Lyrae variables, where many thousands rather than a few tens of cycles are observed, and where the character of the *O–C* diagrams usually rules out such cycle-to-cycle noise as a major contributor to deviations from zero in the *O–C* diagram. The statistical nature of RR Lyrae period changes has been investigated by Balazs-Detre and Detre (1965).

How should the predicted evolutionary period changes of RR Lyrae stars reveal themselves in the *O–C* diagram? In some cases, the evolutionary changes may still be too small to detect after a few tens of years. In those cases, the *O–C* diagram should be essentially a straight line. However, in other stars, especially in those evolving to the red toward the end of their horizontal branch lifetimes, the evolutionary period changes, while still small, ought to be large enough to detect. In those cases, the *O–C* diagram should be well represented by a parabolic curve, which should indicate a slow rate of period increase. A few stars might be caught in the interval of core instability,

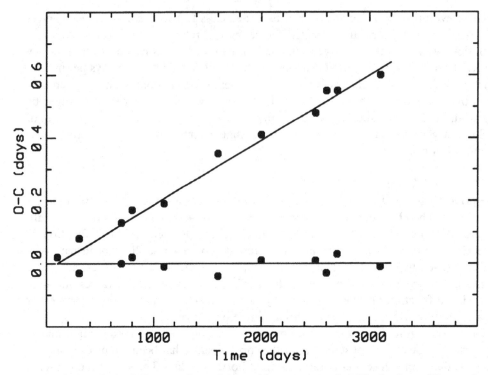

Figure 5.2. *O–C* diagrams for variables with constant period. The horizontal line at *O–C* = 0.0 will result if the period is constant and the correct period and initial epoch of maximum light have been adopted. The sloping line will result if too short a period has been adopted in calculating times of maximum light.

and could show period increases or decreases much larger than $dP/dt = 0.025$ days per million years, but such cases, if they occur at all, would be rare.

The observational picture is quite different.

5.1.3   *The period change noise*

Even before detailed models of horizontal branch evolution were calculated, hopes had dimmed that observed period changes could be directly related to evolution in the HR diagram. Prager (1939), commenting on nearly half a century of observations of three field RR Lyrae stars, noted that their periods did not change at a constant rate, but were better modeled assuming abrupt period changes. Abruptness, of course, does not accord with a model in which period change results from smooth evolution.

Period changes have now been searched for in many field and cluster RR Lyrae stars, in some cases with observations spanning seven or eight decades. Some stars appear to be unchanged in period over that span; others have undergone multiple changes in period. The period changes are small compared to the length of the period itself but because many cycles have gone by, the cumulative effects on the *O–C* diagram can be large. In some stars the observations can be well fit by assuming that the period is changing at a constant rate (parabolic *O–C* diagram), but many others follow Prager's stars in seeming to have abrupt changes in period. When fit by a parabola, the derived rates of period change are usually in the range −1 to +1 days per

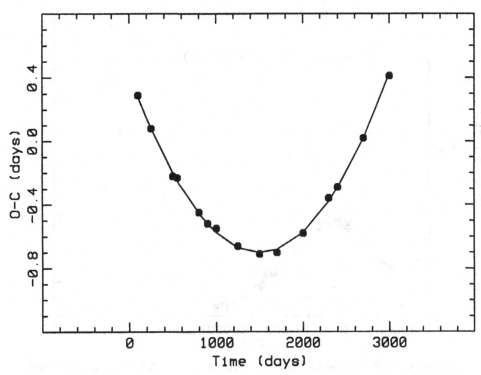

Figure 5.3 If the period of a variable is changing at a constant, relatively slow, rate, the resultant *O–C* diagram is nearly a parabola, as shown for an increasing period.

million years, or a change of less than 10 seconds during the 50–100 years in which the stars have been under observation. A few stars do, however, exhibit period changes several times greater than these limits. Some stars have exhibited both an increase and a decrease in period, as can be seen in the *O–C* diagram of RR Gem in figure 5.4. As we have noted, stellar evolution is not expected to act so quickly as to cause period changes in both directions in so short a span as a century. Moreover, the sizes of the observed period changes are often much larger than expected from theory, even were many of the stars approaching the end of their horizontal branch lives. These observations point to the presence of a 'noise' in the period changes of RR Lyrae stars which overlies any evolutionary period change.

Because of the period change noise, it is impossible to attribute the period changes observed in any individual RR Lyrae star to evolution alone. If there remains hope of soon detecting evolution in RR Lyrae stars by observing their period changes, it must be sought in statistical analyses of the period changes of many RR Lyrae stars.

Before discussing the observations in more detail, we consider whether the theoretical scenario presented in §5.1.1 can be modified to incorporate noisy period changes. Attempts to do this have been made, the most detailed being Sweigart and Renzini's (1979) proposal that the period change noise is a result of random mixing events associated with the semiconvective zone of the stellar core. They noted that changes in the internal structure of an RR Lyrae star arise in two ways: through the gradual change in core composition caused by nuclear burning, but also through a

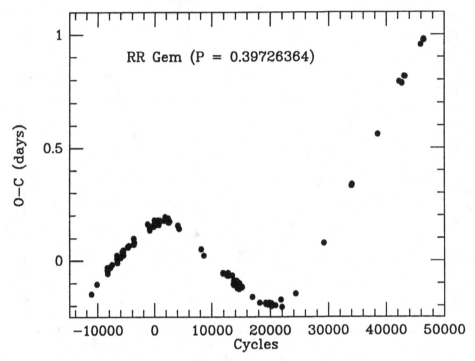

Figure 5.4 *O–C* diagram for times of maximum light of RR Gem, after Tsesevich (1966)

composition redistribution in the deep interior caused by convective overshooting and the formation of a semiconvective zone. Whereas this composition redistribution is assumed in standard calculations to proceed smoothly on the same timescale as nuclear burning, Sweigart and Renzini pointed out that the two need not be perfectly coupled at every instant of time. They proposed that in actual stars the composition redistribution is a discontinuous process. There might then be many mixing events, each slightly altering the internal structure of the star, and each leading to a change in period.

Although a detailed understanding of mixing in the semiconvective zone still eludes current theory, Sweigart and Renzini showed that different types of mixing events can account for both period increases and decreases. They furthermore calculated that the observed timescale of period changes is in at least rough agreement with the mixing event hypothesis. In their view, nuclear burning would be expected to produce a slow rate of period change which would be essentially unvarying over the span of a century. The mixing events, on the other hand, if they took place quickly would produce period changes which would seem abrupt. Mixing events on a longer timescale might produce period changes which, for a century, would seem constant in rate. Of course, over a sufficiently long period of time, the period changes due to nuclear burning and discrete mixing events at the semiconvective zone must average to the value predicted by those evolutionary calculations which assume that the composition redistribution takes place smoothly.

Stothers (1980) has proposed a different origin for the period change noise, invoking hydromagnetic effects in RR Lyrae stars. As is often the case with phenomena involving stellar magnetism, the details of how hydromagnetic effects might produce the period change noise have yet to be worked out. Stothers argued, nonetheless, that the magnitudes of the timescales and the sizes of the period changes which might be associated with magnetic cycles in RR Lyrae stars are not unreasonable when compared to the observations.

More recently, Dearborn et al. (1990) have suggested that exotic particles, such as WIMPS (weakly interacting massive particles), might produce changes in the pulsation periods of RR Lyrae stars. The existence of these particles has been postulated to explain the apparent existence of large amounts of dark matter in galaxies. It has been further supposed that these particles might become trapped within stars, where they can provide a new mechanism of energy transfer. In the case of horizontal branch stars, Dearborn et al. proposed that WIMPS might cause thermal pulses to occur on the Kelvin–Helmholtz timescale for the stellar core. These thermal pulses would produce period changes. However, the timescale calculated for these changes, on the order of 10^6 years, is too long to explain the shorter term period noise.

Coutts (1971) identified another possible cause for apparent period changes in RR Lyrae variables: light travel time effects because of motion in a binary star system. Motion of an RR Lyrae star in a binary could introduce a long term periodic oscillation in the O–C diagram for time of maximum light, as the star alternately approaches and recedes from us. It is difficult, however, to build a case for the binary nature of any RR Lyrae star from O–C data alone, and for comparatively few variables does the binary hypothesis by itself provide an adequate explanation of the period change behavior. Saha and White (1990) have, however, argued on the basis of O–C data and radial velocity observations that TU UMa may be binary.

Koopmann et al. (1993) investigated whether mass loss might alter the theoretical period changes of RR Lyrae stars. They calculated synthetic horizontal branches under the assumption that horizontal branch stars lost mass when they were in the instability strip, that is, when they were RR Lyrae stars. They concluded that the predicted rates of period change were not significantly changed for mass loss rates as high as $10^{-9} M_\odot/$year.

5.1.4 Observed period changes of field RR Lyrae stars

Thanks to the work of many observers at many different observatories over a long interval of time, period changes have been observed for numerous RR Lyrae variables of the general galactic field. Among the contributors to this work have been the Konkoly Observatory, Hungary, the Maria Mitchell Observatory, United States, the Odessa Astronomical Observatory, Ukraine, and the RR Lyrae Committee of the American Association of Variable Star Observers, to mention only a few. For many stars period changes are listed in the notes to the *General Catalogue of Variable Stars*.

Tsesevich (1966, 1972) has given a detailed account of the period changes of many field RR Lyrae stars, mostly of type RRab. He found that some generalizations were possible, despite the fact that the observed period changes differed greatly in size and character from star-to-star. In characterizing the observed period changes, Tsesevich found, when sufficient numbers of accurate observations were available, that only rarely could the O–C diagram of times of maximum be perfectly fit by a parabolic formula. Usually, the period changes revealed in the O–C diagram were more abrupt than a perfect parabola would allow.

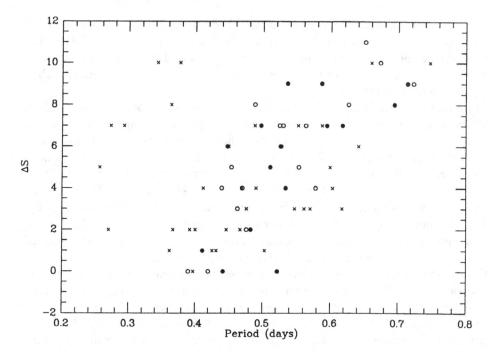

Figure 5.5 Period changes of field RR Lyrae stars as a function of period and metallicity. Period changes have been taken from Tsesevich (1966). Symbols are explained in the text.

Tsesevich also investigated the relationship of period change to metallicity, as determined by Preston's index ΔS (§4.2). He found that, at a given period, the RRab stars of low ΔS (high metallicity) tended to have smaller period changes than those of high ΔS (low metallicity), but that some stars of changing period were found at all metallicities. This is shown in figure 5.5, an updated version of a similar diagram in Tsesevich (1972). Here RR Lyraes with period changes reported by Tsesevich are plotted in the ΔS versus period diagram. The ΔS values were taken from Preston (1959), Butler (1975), and Kemper (1982), or were estimated from the photometric observations of Lub (1979). Recall that, according to Butler (1975), [Fe/H] $= -0.16\Delta S - 0.23$.

In figure 5.5, variables with roughly constant periods are plotted as crosses, those with predominantly increasing periods as open circles, and those with mainly decreasing periods as filled circles. A few variables with ill-defined or very irregular *O–C* diagrams were not plotted. Were the period changes due to evolution alone, we might speculate that the high metallicity RRab's tend to have small period changes because they are closer to their zero-age horizontal branch positions in the HR diagram than are those of low metallicity, but the existence of the period change noise confuses the question. Above and to the left of the RRab stars in figure 5.5 lies a sequence of RRc stars having nearly constant periods. However, Tsesevich reported on period changes for only a few RRc stars and no conclusions can be drawn from this sequence. Many RRc stars with variable period are known in globular clusters and Tsesevich noted the existence of several RRc stars with irregular lightcurves which could not be plotted in figure 5.5.

5.1.5   Period changes of RR Lyrae stars in globular clusters

The RR Lyrae-rich globular clusters probably afford us our best opportunity to directly measure RR Lyrae evolution through the HR diagram. One might hope that, if a cluster has many RR Lyrae stars, and if photometry of those variables were obtained over a long enough interval of time, then the period change noise might be averaged so as to reveal true evolutionary period changes. Indeed, there is evidence that this may have been achieved for the RR Lyraes of one or two globular clusters.

Since Martin's (1938) study of period changes of RR Lyrae stars in ω Centauri, similar investigations have been carried out for many globular clusters. Table 5.1 summarizes recent studies of RR Lyrae period changes, updating a similar table in Stagg and Wehlau (1980). Clusters have been included for which period changes have been determined for ten or more RR Lyrae stars over a span of 30 or more years. All of the clusters belong to the Galaxy, except for NGC 2257, which is an old star cluster of the Large Magellanic Cloud. Column (1) identifies each cluster and column (2) gives its Oosterhoff group. The columns headed $N+$, $N-$, Ncst, and Nirr give the number of RR Lyraes with increasing, decreasing, constant, and irregular periods, respectively. It has not been possible to use uniform criteria for all studies in assigning RR Lyraes to these categories, so that the numbers in these columns should only be regarded as indicative. The period change rates of globular cluster RR Lyrae stars have often been characterized by the quantity β, the rate of period change determined by fitting a parabola to the O–C diagram. Different investigators have used different units for β, a

Table 5.1. *Period changes of RR Lyrae stars in globular clusters*

Cluster	Oo	$N+$	$N-$	Ncst	Nirr	β	ΔT	Source
M3	I	32	33	20	25	0.00	65	Szeidl 1965
M4	I	2	7	19	4	—	77	Sujarkova and Shugarov 1981
M5	I	20	12	18	16	0.00	71	Coutts and Sawyer Hogg 1969
M5	I	27	14	9	—	—	70	Kukarkin and Kukarkina 1973
M5	I	4	2	4	—	—	91	Storm et al. 1991
M14	I	12	14	9	—	−0.02	42	Wehlau et al. 1975
M28	I	2	2	7	—	0.06	43	Wehlau et al. 1986
M107	I	4	5	12	1	0.00	35	Coutts and Sawyer Hogg 1971
M107	I	3	10	6	—	—	35	Gryzunova 1979
NGC6934	I	6	11	31	2	0.00	47	Stagg and Wehlau 1980
NGC7006	I	18	23	5	—	−0.04	53	Wehlau et al. 1992
M15	II	16	6	10	4	0.04	82	Smith and Sandage 1981
M15	II	22	5	—	7	—	85	Gordenko et al. 1984
M15	II	22	10	4	—	0.03	96	Silbermann and Smith 1993
M22	II	8	4	1	4	0.04	82	Wehlau and Sawyer Hogg 1978
M53	II	8	9	4	—	0.00	53	Goranskij 1976
M68	II	6	4	13	—	—	41	Clement et al. 1993
M92	II	2	4	1	3	—	49	Kukarkin and Kukarkina 1980
ω Cen	II	34	9	—	—	0.11	74	Belserene 1964; 1973
NGC5466	II	5	7	8	—	0.00	45	Gryzunova 1972
NGC2257	I-II	4	11	19	1	0.00	31	Nemec et al. 1985
NGC2257	I-II	5	10	15	—	0.00	37	Walker 1989b

common choice being units of days per million years, which we shall adopt. Where the investigators have tabulated β values, the median β for the RR Lyraes is given in column (7) of the table. It must be emphasized, however, that β is merely a convenient shorthand. The actual *O–C* diagrams for cluster RR Lyrae stars, as for their field counterparts, can differ greatly from parabolas (figure 5.6). The *O–C* diagrams themselves must be examined if one is to appreciate the wide range of observed period

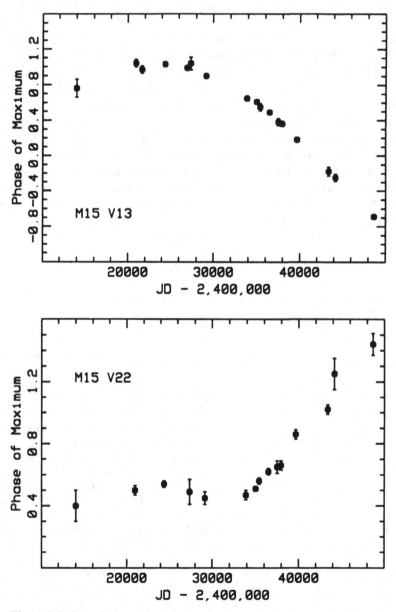

Figure 5.6 Plots of phase of maximum versus Julian Date for two RR Lyrae stars in M15. Observations spanning nearly a century show a period decrease for V13 and a period increase for V22. After Silbermann and Smith (1993).

change behavior. The median β is given since the mean value of β can be strongly affected by a few stars with large values of $|\beta|$. Nevertheless, these outliers cannot be ignored, for it may be that they represent important but brief episodes of period change in the life of a horizontal branch star. The column headed ΔT gives the time interval spanned by those observations used in determining period changes.

A few generalizations are possible from table 5.1. RR Lyraes in the Oosterhoff I clusters tend to have small median period changes and roughly equal numbers of stars with increasing and decreasing periods. The situation may be different for the Oosterhoff II clusters. More stars seem to be increasing in period than decreasing in ω Centauri, and the same may be true to a lesser degree in M22 and M15 (but see Barlai 1984). This trend was noticed by Stagg and Wehlau (1980). It is possible that, at least in ω Cen, the excess of increasing periods reflects RR Lyrae evolution, with most ω Cen RR Lyraes evolving from blue to red through the instability strip. Szeidl (1975) has also suggested that the character of the O–C diagrams might differ from cluster to cluster, with a higher proportion of O–C diagrams in M3 being well-fit by parabolas than is the case in M15.

There has been no unanimity among the various theoretical predictions concerning the direction of RR Lyrae evolution in globular clusters. Van Albada and Baker (1973) in their hysteresis model of the Oosterhoff dichotomy (§3.5) proposed that evolution is primarily to the blue for RR Lyraes in Oosterhoff I clusters and to the red in clusters of type Oosterhoff II. Thus, Oosterhoff I RR Lyraes should show mostly decreasing periods and Oosterhoff II RR Lyraes increasing periods. While it is true that the observations show more increasing periods among the RR Lyraes of the Oosterhoff II clusters, the Oosterhoff I RR Lyraes seem to have roughly equal numbers of increasing and decreasing periods. Cox (1980), on the other hand, argued that most globular cluster RR Lyraes are evolving to the blue, so that decreasing periods ought to predominate in both clusters of Oosterhoff I and Oosterhoff II.

Lee, Demarque, and Zinn (1990) have suggested that most RR Lyrae stars in Oosterhoff I clusters are located near their zero-age horizontal branch positions in the HR diagram, while most of those in Oosterhoff II clusters have evolved into the instability strip from initial locations on the blue horizontal branch. It is tempting to see the period change results as reflecting this scenario. The Oosterhoff I RR Lyraes, being near their zero-age horizontal branch locations, would be expected to show small evolutionary period changes. The Oosterhoff II RR Lyraes, moving from blue to red through the HR diagram toward the asymptotic red giant branch, would be expected to have mainly increasing evolutionary period changes of the greater size appropriate to RR Lyrae stars in the later stages of their horizontal branch lives.

Predicted period changes for this scenario by Lee (1991b) are compared with observations of period changes in globular clusters in figure 5.7. In this figure, median values of period change are plotted against a measure of horizontal branch type. B, V, and R in this case stand for the number of horizontal branch stars to the blue of the instability strip, in the instability strip (RR Lyraes), and to the red of the instability strip, respectively. The solid line indicates the theoretical prediction of Lee (1991b). There is indeed a general agreement between theory and observation, but the evidence for a period increase for RR Lyraes in blue horizontal branch clusters hangs most strongly on ω Centauri. The data at hand are broadly consistent with Lee's predictions, but are not adequate to provide confirmation.

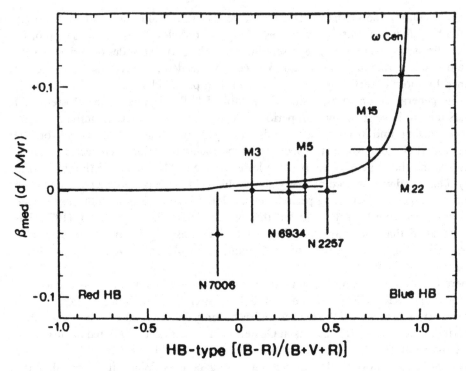

Figure 5.7 Median period changes for RR Lyraes in several globular clusters are plotted against a measure of horizontal branch type. The solid line indicates the theoretical period changes calculated by Lee (1991b). From Wehlau et al. (1992).

NGC 7006 is a classic second parameter globular cluster, having an unusually red horizontal branch for its metallicity. It is therefore interesting that, of all well observed globular clusters, NGC 7006 has the most negative median rate of period change. Wehlau et al. (1992) also noted that there was evidence for a radial gradient in period changes among the NGC 7006 RR Lyraes, with β values being more negative in the outer portion of the cluster. The LMC cluster NGC 2257 also has a strongly negative mean rate of period change, but its median β value is near 0.00.

Photometric data for RR Lyraes in several of the globular clusters now extend more than seven decades into the past. Yet, only now have enough stars been observed for a sufficiently long time in enough clusters for astronomers to begin to seriously test theories of RR Lyrae evolution. To yield conclusive results, this collaboration between astronomers of the past and present must be continued into the future. It is important that all RR Lyrae-rich globular clusters be observed regularly, both to increase the intervals over which period changes can be studied, and also to insure that there are no gaps in the observational record which might lead to cycle count ambiguities and uncertain *O–C* diagrams.

5.2   The Blazhko effect

The lightcurves of many RR Lyrae stars repeat precisely or nearly so from cycle-to-cycle. Others do not, but instead have lightcurves which change form significantly on a relatively short timescale. For some of these stars, the lightcurve changes appear to be irregular, but in many instances the lightcurve variability can be attributed to the presence of a second periodicity. RR Lyrae variables which exhibit secondary periodicities can be divided into two main groups: the so-called RRd stars which show a mixture of fundamental and first overtone radial pulsation modes (see §5.3) and the stars which show the classical Blazhko effect.

5.2.1 Observations

The Blazhko effect manifests as a periodic modulation of the lightcurve shape on a timescale typically of tens of days. It was first noticed by S. Blazhko (1907) who found that a single period did not well describe times of maximum of the RR Lyrae star RW Dra. Blazhko found that he had to introduce a 41.6 day cyclic variation in the period of this star to satisfy the observations. Soon thereafter, Sperra (1910) reported similar short-term changes in the lightcurves of SU Dra and SW Dra, although curiously these have not been seen in more recent observations (Szeidl 1976). It was Harlow Shapley (1916) who demonstrated that RR Lyrae itself shows a secondary Blazhko periodicity. He found that the lightcurve of RR Lyrae changed form over a cycle of about 40 days, a period which later work has refined to 40.8 days. The differences in the lightcurve of RR Lyrae at two points in its 41 day cycle are illustrated in figure 5.8. Notice that the difference in brightness at maximum light is more pronounced than at minimum. This is often, but not always, the case for Blazhko effect stars. The change in lightcurve shape often results in a periodic change in the times of maximum light, so that the Blazhko effect is evident in the $O–C$ diagram for times of maximum.

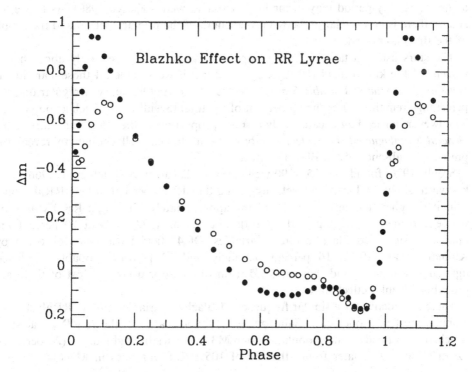

Figure 5.8 The lightcurve at two different points in the 41 day Blazhko cycle of RR Lyrae, based upon photoelectric photometry by Walraven (1949).

Blazhko periods have now been well established for about 40 RR Lyrae stars in the general galactic field, and the existence of the phenomenon has been suspected in many other, less well-observed, field stars. The Blazhko effect is also believed to occur in RR Lyrae stars within globular clusters, though for only a few cluster variables has the length of the secondary cycle actually been determined.

RR Lyrae stars for which Blazhko periods have been determined are listed in table 5.2, a somewhat revised version of Szeidl's (1976) table 1 and Szeidl's (1988) table 3. Sources for the primary and secondary periods may be found in Szeidl's original tables, except for AH Cam (Goranskij 1989 and Smith et al. 1994), and SW Boo (Taylor 1977). Values followed by a colon are uncertain. Where it is known, the metallicity index ΔS is also listed. For RRab stars, the ΔS values from the compilation of Blanco (1992) were used when they were available, but values for the RRc stars are from Kemper (1982). ΔS values for the globular cluster stars are taken from Butler (1975) and Smith (1984b), and are the mean ΔS values for RR Lyraes in a given cluster. Only three of the stars in table 5.2 are RRc's; the remainder are of type RRab. While the paucity of RRc variables known to show the Blazhko effect may be at least in part a selection bias (lightcurve variability may be less pronounced and harder to identify in the low amplitude RRc's), it probably cannot be attributed entirely to such a bias. In fact, it is not certain whether any RRc stars exhibit the classical Blazhko effect. The existence of the secondary periodicity of one of the three RRc variables, RU Psc, has been questioned (Mendes de Oliveira and Nemec 1988).

The shortest known Blazhko cycle is the 10.9 days reported for AH Cam. The longest is the approximately 533 day period reported for RS Boo (Oosterhoff 1946, but a shorter 62 day period may occur in that star as well – Kanyo 1980). As figure 5.9 shows, there is no correlation between the length of the primary period and the length of the Blazhko period.

The stars listed in table 5.2 are comparatively few in number, only about half a percent of the known field RR Lyrae stars. Even if we include all those stars in the *General Catalogue of Variable Stars* suspected of showing lightcurve irregularities, the percentage remains at less than 5 percent of the total (Szeidl 1988). Such a comparison is, however, misleading because only a small proportion of the RR Lyrae stars in the *General Catalogue of Variable Stars* have been studied well enough to reveal the presence or absence of the Blazhko effect.

Szeidl (1976) found that 15 of 90 well-observed RRab stars (Fitch et al. 1966) were known to show the Blazhko effect, suggesting that 15–20 percent of field RRab's may exhibit the phenomenon. Similarly, of the approximately 150 bright RR Lyrae stars whose ephemerides are reported by Firmaniuk et al. (1988), about 20 percent are known or suspected Blazhko effect variables. Of 42 field RRab stars observed by Kinman et al. (1984), 16 percent definitely and 12 percent probably displayed lightcurve scatter beyond that expected from observational error. Most of these are probably Blazhko effect stars.

There is evidence for a similar frequency of Blazhko variables among RRab stars in globular clusters. Roberts and Sandage (1955) found that 22 of the 78 RR Lyrae stars which they studied in the globular cluster M3 had variable lightcurves (28 percent). Szeidl (1965, 1976) later found that 36 of 105 RRab variables in M3 (34 percent)

Table 5.2. *RR Lyrae stars with known Blazhko cycles*

Variable	Type	P (days)	P_{Bl} (days)	ΔS
AH Cam	ab	0.369	10.9	−1:
RS Boo	ab	0.377	533	1.6
RR Gem	ab	0.397	37	1.6
MW Lyr	ab	0.398	33.3	—
DM Cyg	ab	0.420	26.0	1.7
SW And	ab	0.442	36.8	−0.4
RW Dra	ab	0.443	41.6	6.4
RV Cap	ab	0.448	225.5	6.9
BI Cen	ab	0.453	70:	2.3
TU Com	ab	0.461	75	—
XZ Cyg	ab	0.467	57.3	6.2
RV Uma	ab	0.468	90.1	4.4
AR Her	ab	0.470	31.6	6.3
XZ Dra	ab	0.476	76	3.3
RY Col	ab	0.479	90:	3
M5 V14	ab	0.487	75.0	5.3
X Ret	ab	0.492	45:	3
V674 Cen	ab	0.494	29.5:	—
M5 V63	ab	0.498	146.8	5.3
KM Lyr	ab	0.500	30	—
M3 V5	ab	0.506	194.6	8.4
RZ Lyr	ab	0.511	116.7	9.3
SW Boo	ab	0.514	13.0	—
V434 Her	ab	0.514	26.1	—
SW Psc	ab	0.521	34.5	—
Y LMi	ab	0.524	33.4	—
M5 V2	ab	0.526	132:	5.3
M53 V30	ab	0.535	37.0	10.1
SZ Hya	ab	0.537	25.8	6
UV Oct	ab	0.543	80:	9.3
V788 Oph	ab	0.547	115:	—
RW Cnc	ab	0.547	87	8.7
AD UMa	ab	0.548	35–40	—
TT Cnc	ab	0.563	89	7.5
RR Lyr	ab	0.567	40.8	6.1
V829 Oph	ab	0.569	165:	—
AR Ser	ab	0.575	105	6
WY Dra	ab	0.589	14.3	—
DL Her	ab	0.572	33.6	6
V365 Her	ab	0.613	40.6	—
ST Boo	ab	0.622	284	9.0
BH Peg	ab	0.641	39.8	5.4
Z CVn	ab	0.654	22.7	8:
TV Boo	c	0.313	33.5	12
BV Aqr	c	0.364	11.6	8
RU Psc	c	0.390	28.8?	9

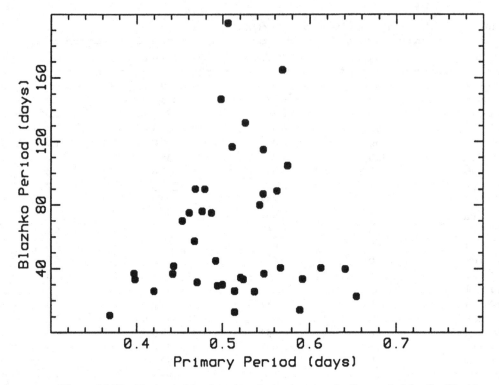

Figure 5.9 Blazhko period is plotted against primary pulsation period for the Blazhko effect RRab stars of table 5.2.

showed lightcurve variability, while in M15 the percentage of RR Lyraes with variable lightcurves was about 25 percent (Barlai, reported in Szeidl 1976). Smith (1981) estimated that about 25 percent of the RR Lyrae stars in the globular clusters M3, M5, M15, and ω Cen, and in the Draco dwarf spheroidal galaxy, showed lightcurve variability. It is now known that not all of these stars show the classical Blazhko effect: some are RRd type variables. This is especially true of stars formerly classified as RRc variables with inconstant lightcurves. Nevertheless, the data are consistent with the estimate that 20 percent or more of globular cluster RRab variables are Blazhko effect stars, a rate comparable to that seen among well-observed field RRab variables.

Many of the M3 RRab variables with varying lightcurves probably show the classical Blazhko effect although secondary periods have for the most part not been determined for them. Szeidl (1976, 1988) found that when these variables were at their greatest light amplitude, they fell approximately on the curve of amplitude versus period as defined by stars with regular lightcurves. Szeidl found this also to be true of a group of well-observed field RRab stars. Thus, the Blazhko effect, when it does act, reduces rather than extends the normal height of maximum (see also, Teays 1993).

Szeidl (1976) found that the Blazhko effect was more common among metal deficient RRab variables than among those of near solar abundance. He estimated that roughly 10 percent of RRab stars with $\Delta S = 0$–2 showed the effect, whereas the proportion rose to 20 percent for RRab's with $\Delta S = 3$–5, and to 30 percent for $\Delta S = 6$–10. Other results are more ambiguous. In table 5.3 the field RRab stars in table 5.2 are divided

Table 5.3. *Blazhko effect as a function of metallicity*

ΔS	Blazhko RRab stars		Bright RRab stars	
	No.	%	No.	%
0–2	5	21	18	24
3–5	5	21	21	28
6–11	14	58	35	47

into three metallicity groups. For comparison, those bright RRab stars in Firmaniuk et al. (1988) with ΔS values by Preston (1959) are also listed. No trend in prevalence of the Blazhko effect as a function of metallicity is evident in this comparison.

Preston, Smak, and Paczynski (1965) carried out extensive simultaneous photometric and spectroscopic observations of RR Lyrae itself, concentrating particularly on phases of rising light. They found the 41 day cycle strongly present in 1962 and 1964, but nearly absent during 1963. RR Lyrae, like many other RRa variables, shows brief hydrogen line emission in its spectrum at certain phases during rising light. Struve (1947) had earlier suspected that the strength of this emission varied during the 41 day cycle, a suspicion which Preston et al. were able to confirm. The amount of emission was greatest when RR Lyrae was at a phase in its Blazhko cycle which gave the greatest light amplitude. Preston, Smak, and Paczynski also observed that the character of doubling of hydrogen and metallic lines changed during the 41 day cycle. Line emission and line doubling in the spectra of RR Lyrae stars have been attributed to the presence of a shock wave during rising light (see chapter 4). The observations of Preston, Smak, and Paczynski therefore indicated to them that the character of this shock was modified during the Blazhko cycle. They suspected that a critical level of shockwave formation moved higher and lower in the atmosphere of RR Lyrae during the course of the secondary period. Struve and Blaauw (1948) and Preston et al. (1965) found that the Blazhko effect in RR Lyrae was reflected in its radial velocity curve as well as in its lightcurve.

A pronounced characteristic of the Blazhko effect is its irregularity. In at least some instances the behavior of the star does not repeat precisely at all Blazhko cycles. In some stars, as with RR Lyrae itself, the Blazhko effect nearly vanishes some years, but is very strong in others. The Blazhko effect in RR Gem appears to have ceased about 1940 (Detre 1969; Szeidl 1976). In the metal-rich RRab star SW And the Blazhko effect now manifests as a periodic change in the form of a hump on the ascending branch of the lightcurve, but may have been more pronounced before 1956 (Szeidl 1976). Many Blazhko effect stars have been observed to undergo changes in primary period, often of a complicated character. These changes in the primary period are often reflected in changes of the Blazhko period. This is shown for RW Dra in figure 5.10. The main period and the Blazhko period usually change in different directions, but this may not be invariably the case.

Some researchers have reported third periods for Blazhko effect stars, but the long term existence of these additional periodicities is in many cases uncertain. As noted above, RS Boo may have a 62 day as well as a 533 day period. Likewise, RW Cnc, XZ

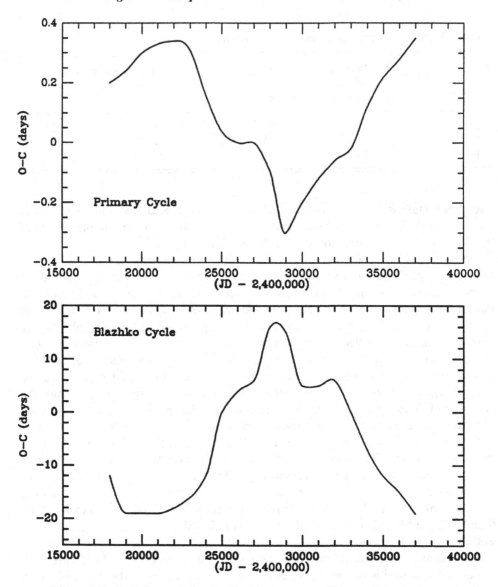

Figure 5.10 *O–C* diagrams for the Blazhko cycle and the primary pulsation cycle of RW Dra are compared. Note that the two *O–C* diagrams usually show opposite behavior. After Tsesevich (1966).

Cyg, RW Dra, AR Her, Y LMI, and RR Lyr itself may show third periods, usually near 100 days in length (table 4 of Szeidl 1988).

In a few cases, very long period cycles in the amplitude of the Blazhko effect seem to exist. There is a four year cycle for RR Lyrae (Detre and Szeidl 1973), a nine year cycle for XZ Cyg (Klepikova 1958), a seven year cycle for RV UMa (Kanyo 1975), a seven year cycle for RW Dra (Szeidl 1988), and an eight year cycle for Y LMi (Szeidl 1988). Detre and Szeidl found that the start of a new four year cycle for RR Lyrae was accompanied by an abrupt phase shift of about ten days in the Blazhko cycle.

Hoffmeister (Hoffmeister et al. 1985) reported a tendency for increased scatter on the descending branch of the lightcurves of many RR Lyrae stars, near phase 0.3 after maximum. This scatter presumably arises from the lightcurves not repeating precisely from cycle-to-cycle. Hoffmeister did not believe that this irregularity in the lightcurve was the classical Blazhko effect. The prevalence of this type of irregularity, and its nature, remain to be explored.

5.2.2 *Explanations of the Blazhko effect*

Stellingwerf (1976), in a review of multiperiodic RR Lyrae stars, listed six mechanisms which might create the Blazhko effect:

1. Resonance effects in radial modes
2. Resonances involving nonradial modes (Ledoux 1951; Cox 1993a)
3. Splitting of radial modes caused by nonadiabatic effects (Ledoux 1963)
4. Tidal effects in binary systems (Fitch 1967)
5. Oblique rotator effects (Balazs-Detre 1959; Preston 1964)
6. Magnetic cycle effects (Detre and Szeidl 1973)

Most speculation has centered on two possibilities. These are (1) that the Blazhko effect is the consequence of some type of mixing of pulsational modes and (2) that the Blazhko effect is related to magnetic cycles in the stars, perhaps coupled with rotation. Evidence can be marshalled to support either hypothesis, but both have weaknesses and neither can yet be said to be established.

The idea that the Blazhko effect represented some sort of interference between the primary pulsation mode (usually the fundamental) and another mode is an old one, but early attempts to establish this were unsuccessful (Kluyver 1936). It is clear from the observations that the Blazhko effect cannot be explained as a simple linear superposition of two pulsations. If the Blazhko effect is a result of modal interference, the two oscillations must combine in a nonlinear fashion.

Tsesevich (1975) believed that this interference was more clearly represented in the radial velocity curves of the Blazhko stars than in the lightcurves. He proposed that the Blazhko period be regarded as a beat period. In that case, if P_B is the Blazhko cycle and P_1 is the primary pulsation period, then the period of the perturbing oscillation, P_2, is given by $1/P_B = 1/P_2 - 1/P_1$, assuming P_1 to be the longer of the two. Because the Blazhko cycle is equal to tens of the primary pulsation period, the period of the perturbation must be nearly equal to that of the primary pulsation mode. Thus, unlike the RRd stars, the Blazhko effect stars cannot be explained by mixing of the fundamental and first overtone radial modes, which would produce a period ratio of roughly 3/4.

The hypothesis that the Blazhko effect is some sort of modal mixing nonetheless finds indirect support in the period distributions of those globular cluster RR Lyrae stars which have variable lightcurves. Smith (1981) and Nemec (1985b) pointed out that in several clusters and in the Draco dwarf spheroidal galaxy such stars tend, if they are RRc stars, to occur near the long period end of the period distribution of RRc variables. In M15, at least, this is clearly because the so-called variable lightcurve RRc's are actually RRd variables, showing a mixture of fundamental and first overtone modes. A related but opposite tendency is seen among the RRab stars. Among these stars, those with variable lightcurves tend to have shorter periods than do the RRab variables as a whole. It is tempting, then, to believe that this, too, must reflect some mixing of pulsation modes, though this cannot be mixing of the fundamental and first overtone modes. Smith (1981) suggested that the Blazhko effect RR Lyrae stars in the general field were likewise deficient in long period variables, a result confirmed by Gloria (1990). Szeidl (1988) found that there are no unquestionable Blazhko effect stars with period longer than 0.7 days among the field RR Lyrae stars.

Although suggestions have been made that the perturbing mode is nonradial, usually the cause of the Blazhko cycle has been sought in interference by another radial pulsation mode. Borkowski (1980) attempted to explain the Blazhko effect in AR Her as the superposition of the fundamental mode (of frequency ν_0) and another mode of frequency $\nu = 2\nu_0 + \nu_B$, where ν_B is the frequency of the Blazhko cycle. The period of the interfering mode would thus be not nearly equal to that of the fundamental, but would be close to half its value. Borkowski identified the perturbing mode as the second or third overtone. Although Borkowski had some success in describing the light variations of AR Her under this hypothesis, the beat mass which he obtained for AR Her, about 1 $M_\odot$, is larger than found for RR Lyrae stars either by stellar evolution theory or by analyses of RRd variables, which give values of 0.6–0.8 $M_\odot$.

Moskalik (1986) investigated the effects on the amplitude stability of RR Lyrae stars of a 2:1 resonance between the radial fundamental mode and the damped third overtone. He found that, under certain conditions, no stable equilibrium solution existed, with the result that the amplitude underwent a periodic modulation, mimicking the behavior of the Blazhko effect. Moskalik went on to suggest that there might be a connection between the classical Blazhko effect and the RRd variables. He suggested that, because the 2:1 resonance lowered the amplitude of the fundamental mode, it might also affect the nonlinear stability of the first overtone mode: the same resonance which produces the Blazhko effect might, under slightly different circumstances, result in RRd pulsation. If so, it might be reasonable to expect that RRd and Blazhko effect stars would occur at nearly the same position in the HR diagram and at similar values of fundamental period.

Though Moskalik's hypothesis is promising, it does not seem to account for RRc variables like BV Aqr and TV Boo, for which Blazhko cycles have been reported. It is therefore important in this connection to confirm that the lightcurve changes of these stars are of the classical Blazhko variety and additional observations of the stars would prove valuable.

The hypothesis that the Blazhko phenomenon is a result of modal interference thus has some theoretical underpinning and can, at least in very rough terms, account for the period distributions of variable lightcurve RR Lyraes in globular clusters. It is, nonetheless, at present far from being able to describe the full range of Blazhko effect behavior. Observations of period changes of Blazhko effect stars may eventually provide useful tests of the modal interference hypothesis. Any modal interference hypothesis must account for the circumstance that the primary period and the Blazhko period usually show changes of opposite sense.

The alternative view, that the Blazhko phenomenon might be magnetic in origin, also now has a considerable history. Babcock (1955; 1958) searched for the Zeeman effect in the spectrum of RR Lyrae. He reported finding a magnetic field the strength of which changed from +1170 to −1580 gauss. Babcock found the field to change significantly on a timescale of hours, but that longer term variations might also be present. Balazs-Detre (1959) found some evidence of a correlation between phase in the Blazhko cycle and Babcock's determinations of magnetic field strength. Cousens (1983) made a theoretical investigation of an oblique magnetic rotator model for RR Lyrae stars and concluded that it had promise for explaining the Blazhko effect.

However, support for a magnetic origin of the Blazhko effect was weakened when Preston (1967) failed to detect a significant magnetic field in RR Lyrae in spectra obtained in 1963 and 1964. Detre (1969) suggested that Preston's observations were not conclusive because of the variability of the Blazhko effect itself. During 1963, Preston, Smak, and Paczynski (1965) had reported the Blazhko effect to be practically absent.

More recently, Romanov, Udovichenko, and Frolov (1987) used the Soviet 6-m telescope to again search for the Zeeman effect in the RR Lyrae spectrum. They found a magnetic field which varied on both short and long timescales, as Babcock had originally reported. In addition to changes associated with phase in the primary pulsation cycle, they found evidence for a variation with the phase of the Blazhko effect. This is the strongest direct evidence that the Blazhko effect may in part be a magnetic cycle, but the observations of Romanov et al. need to be confirmed.

Szeidl (1976, 1988) has remarked that the 4–10 year cycles in the Blazhko effect observed for several RR Lyrae stars are reminiscent of the 11 year solar cycle, but there is of now no direct evidence associating these with magnetic variations. Clearly, it would be very valuable to have new determinations of magnetic field strength for RR Lyrae variables, though this may be a difficult observational task.

5.3   Double mode RR Lyraes

The realization that the lightcurves of some RRc variables in globular clusters showed scatter beyond that expected from observational error is not a new one. Nevertheless, it was not until the 1980s that particular attention was focused on these stars. Sandage, Katem, and Sandage (1981) in an extensive photometric study of RR Lyrae stars in the globular cluster M15 noticed an abrupt increase in scatter among the light curves of the longest period RRc's (figure 5.11). The amplitudes of these variables were seen to change on a short timescale. Because the period interval for the effect was small (0.390 < P < 0.429 days), and because it occured at the transition period between RRc and RRab stars, Sandage, Katem, and Sandage suggested that the phenomenon was due to mode mixing between the fundamental and first overtone modes.

Mode mixing in RR Lyrae stars had previously been established only in the field star AQ Leonis (Cox, King, and Hodson 1980), but analyses of the lightcurves of the M15 variables confirmed that they were indeed double mode stars (Cox, Hodson, and Clancy 1983; Nemec 1985a). These stars show a combination of first overtone and fundamental radial pulsation modes, as Sandage, Katem, and Sandage had suggested. Nemec termed these double mode variables RR Lyrae stars of type RRd.* RRd variables have assumed particular importance because their double mode nature affords an opportunity to determine their masses largely independently of stellar evolution theory.

---

* The name RRd has also been applied by Diethelm (1983) to metal-poor pulsating stars of period 1–3 days. Diethelm (1986) concluded, however, that the 1–3 day period variables probably differ from the classical RR Lyrae stars in that they are in a post horizontal branch stage of evolution. In light of this, it is perhaps preferable to retain the RRd designation for the double mode stars, which are more clearly associated with the RR Lyraes of types ab and c.

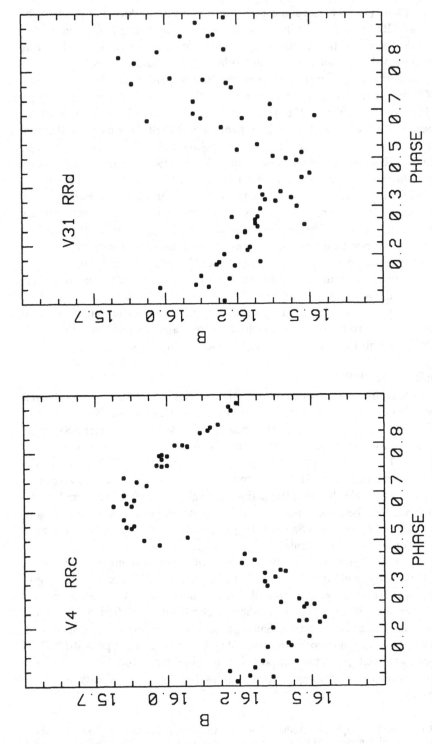

Figure 5.11 Lightcurves of two **RR Lyrae** stars in M15 in which the first overtone mode is dominant. V4 is an RRc star which shows only the first overtone and has little scatter in its lightcurve. V31 is an RRd star. The presence of the secondary fundamental mode oscillation causes a large amount of scatter when the lightcurve of V31 is constructed with the first overtone period alone. Adapted from Sandage, Katem, and Sandage (1981).

5.3.1 Observations of RRd stars

Only a single RR Lyrae of the galactic field, AQ Leonis, has been clearly shown to exhibit RRd behavior (Cox, King, and Hodson 1980), though Clement et al. (1991) have recently identified two other strong field RRd candidates. RRd stars have been discovered in several globular clusters, in the Draco dwarf spheroidal galaxy, and, less certainly, in the Magellanic Clouds (chapter 6). A list of systems with known RRd pulsators is given in table 5.4.

Almost as interesting as the systems in which RRd variables have been found are the systems in which they have been searched for in vain. M5, an RR Lyrae-rich Oosterhoff I cluster like M3, has no known RRd variables (Clement and Nemec 1990), M3 has two, while the Oosterhoff I cluster IC4499 has at least 13. The Oosterhoff II cluster M15 has at least a dozen RRd stars and the Oosterhoff II cluster M68 may have as many as ten. On the other hand, the RR Lyrae-rich Oosterhoff II cluster ω Centauri apparently has none (Nemec et al. 1986). The dwarf spheroidal galaxy Draco contains perhaps ten RRd stars, while none have been discovered in the Ursa Minor dwarf spheroidal system (Nemec et al. 1988). Clement and Nemec (1990) and Clement and Walker (1991) failed to identify RRd variables in several other well observed globular clusters. Any full theoretical explanation of RRd pulsation must account for the absence of RRd variables in these systems, as well as for the presence of RRd stars in the systems of table 5.4. With the single exception of V68 in M3 (Nemec and Clement 1989), the amplitude of the first overtone pulsation exceeds that of the fundamental mode pulsation in all known RRd stars.

As attested by these searches, the proportion of RRd variables in a system is not a fixed fraction of the total number of RR Lyrae stars, nor of the populations of RRc or RRab variables. Clement and Walker (1991) noted that RRd variables seemed to be found only in relatively metal-poor systems. None of the more metal-rich Oosterhoff I globular clusters appear to contain RRd variables. Nevertheless, a very low metal abundance, while perhaps necessary to the production of RRd variables, is clearly not a sufficient condition for their occurrence.

Table 5.4. *Systems known to contain RRd variables*

System	[Fe/H]	no. RRd	Reference
M3	−1.57	2	Goranskij 1981
			Nemec and Clement 1989
M15	−2.15	12	Sandage et al. 1981
			Cox et al. 1983
			Nemec 1985a
			Jurcsik and Barlai 1990
M68	−2.09	9	Andrews 1980
			Clement 1990
			Clement et al. 1993
IC 4499	−1.50	13	Clement et al. 1986
NGC 2419	−2.10	1	Clement and Nemec 1990
NGC 6426	−2.20	1	Clement and Nemec 1990
Draco	−1.7 to −2.6	10	Nemec 1985b
Galactic field	—	1–3	Jerzykiewicz et al. 1982
			Clement et al. 1991

5.3.2   *Interpretation of RRd variables*

There is no satisfactory theory for the existence of RRd pulsators. It might be assumed that RRd stars are in the process of switching pulsation modes from first overtone mode to fundamental mode or vice versa. There is evidence that the RRd phenomenon can cease on a short timescale. Clement et al. (1993) found that V21 in M68 was a double mode pulsator as recently as 1986–1988, but could detect only the first overtone mode in observations of V21 made in 1989–1991. However, Cox, Hodson, and Clancy (1983) and Cox, King, and Hodson (1980) calculated that detectable double mode behavior should extend over no more than a few hundred or, at most, a few thousand years. That is far too short to account for the observed number of RRd stars in clusters like M15. More recent calculations have had only limited success in extending double mode behavior for longer intervals of time, and the problem remains unsolved.

The reason for the greatly different frequencies of RRd variables among otherwise rather similar clusters is also not known. Nemec and Clement (1989), noting that RRd pulsators lie in the transition zone in the HR diagram between RRc and RRab stars, considered whether the width of the transition zone in the HR diagram might play a role in producing these variables. They concluded, however, that there was no evidence for a difference in the widths of the transition zones between M3 and IC4499 or between ω Centauri and M15. Nemec, Linnell Nemec, and Norris (1986) speculated that differences in stellar rotation speeds between clusters could affect the frequency with which RRd's are produced, but this remains an untested possibility. Cox (1987) argued on theoretical grounds that a high helium abundance might promote RRd behavior. If that were the case, those clusters with RRd variables might have horizontal branch stars with higher helium abundances than similar clusters lacking in RRd stars. Wilson and Bowen (1984) suggested that mass loss might cause RRd variables to be trapped in the mode switching region, but this is so far without observational support and still does not explain why such a mechanism would work in some clusters but not others. Smith and Sandage (1981) and Jurcsik and Barlai (1990) used observed period changes of RRd stars in M15 to investigate whether the direction of evolution in the HR diagram might determine if an RR Lyrae becomes an RRd variable. They were unable to reach any definite conclusions on the question because of the difficulty in deducing the long-term direction of evolution from period change observations of M15 variables spanning less than a century.

The RRd stars are important because they have the potential of providing information about the physical properties of RR Lyrae stars independently of much of the usual stellar evolution theory. In particular, stellar pulsation theory can be used to determine the masses of the RRd variables, and hence to compare the results of stellar evolution theory and stellar pulsation theory.

Beginning with the pulsation equation, $Q = P\sqrt{\rho}$, it is seen that the pulsational mass of an RR Lyrae star can be calculated from

$$M/M_\odot = (Q_i/P_i)^2(R/R_\odot)^3 = (Q_i/P_i)^2(L/L_\odot)^{1.5}(T_e/T_\odot)^{-6}$$

where the subscript i indicates the pulsation mode and Q is the usual pulsation constant. Q can be calculated theoretically. Theoretical pulsation calculations show that Q depends mainly on the mass and radius of the star, and less strongly upon other quantities such as effective temperature or chemical composition (Petersen 1990a).

In principle, the above equation can be used to calculate pulsational masses for RR

Lyrae stars which pulsate in only one mode, as well as for doubly periodic RR Lyraes (see, for example, Carney, Storm, and Jones 1992). In practice, while the period is known very exactly, the luminosity and effective temperature are significantly uncertain so that, for single mode stars, the calculated masses provide only a rough check upon stellar evolution theory. Somewhat more accurate results can be obtained differentially, as, for example, when pulsation theory is used to calculate the difference in mass between two RR Lyrae stars in a single globular cluster. However, the most accurate results are possible for double mode RR Lyrae stars. In that case, the two pulsation periods can be known very well and the same luminosity and effective temperature apply to both modes. Petersen (1973) pointed out that masses for double mode variables can be read off a diagram which plots the period ratio, P_1/P_0, against the fundamental period, P_0. This type of diagram is often referred to as a Petersen diagram. Strictly speaking, it is the mass as a function of luminosity which can be determined, though the derived masses are not strongly dependent upon luminosity.

When they were plotted in the Petersen diagram, it became apparent that the RRd variables occupied two distinct locations (figure 5.12). The RRd stars in Oosterhoff II clusters, plus nine of the ten Draco RRd stars, and AQ Leo occupied one region of the Petersen diagram, while the RRd stars in Oosterhoff I clusters and one of the Draco RRd stars occupied a quite different location. Initially, this seemed to imply that the two groups of RRd stars had two distinct masses.

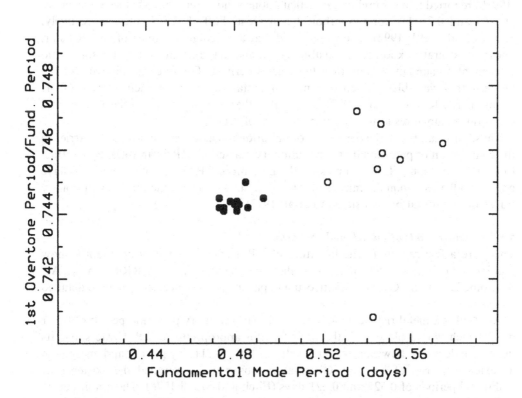

Figure 5.12 RRd variables in the Oosterhoff II globular clusters M15 and M68 (open circles) and the Oosterhoff I globular cluster IC 4499 (filled circles) are plotted in the Petersen diagram. Data are from table 7 of Clement et al. 1993.

Through the 1980s and into the 1990s, most calculations of RRd star masses obtained results similar to those presented by Cox, Hodson, and Clancy (1983). From their location in the Petersen diagram, Cox et al. determined that RRd stars in Oosterhoff II clusters like M15 had masses of about 0.65 $M_\odot$, while those in Oosterhoff I clusters like M3 had smaller masses, near 0.55 $M_\odot$. These results posed a perplexing problem because conventional stellar evolution tracks predicted a higher mass for RR Lyraes (about 0.75 $M_\odot$ in general), and a smaller mass difference between RR Lyrae stars in the two Oosterhoff groups (Simon and Cox 1991).

These troublesome results were based upon pulsational calculations which used opacities determined at Los Alamos. Recently, new opacity calculations at Lawrence Livermore National Laboratory (Iglesias, Rogers, and Wilson 1990) have altered this picture. Although, as of this writing, the full ramifications of these Livermore OPAL opacities for RR Lyrae pulsation have not yet been worked out, the consequences for the masses of RRd variables appear to be important. Kovacs, Buchler, and Marom (1991) and Cox (1993b) recalculated the masses of the RRd stars taking account of the new Livermore opacities, with results substantially different from earlier calculations. Masses for both Oosterhoff I and Oosterhoff II variables appear to be higher than in the earlier calculations. Kovacs et al. concluded that RRd masses could be as high as 0.8–0.9 $M_\odot$, with the Oosterhoff II stars being slightly more massive than the Oosterhoff I stars. Instead of being too low, these revised RRd masses are in some instances now too high compared to stellar evolutionary calculations. However, Cox (1993b) reported that his preliminary calculations yielded the somewhat smaller masses of 0.65 and 0.75 $M_\odot$ for Oosterhoff I and Oosterhoff II RR Lyraes, respectively. Kovacs et al. (1991; 1992) have emphasized that to calculate the mass of an RRd star, one must accurately know its metal abundance, and that deviations from the solar ratio of elemental abundances can affect the masses derived. This is especially true for the Oosterhoff I variables, which are not so metal deficient as their Oosterhoff II counterparts. Kovacs et al. (1992) commented that existing uncertainties in chemical composition alone result in mass uncertainties of $0.1 M_\odot$.

Thus, uncertainties in the derivation of pulsational masses remain a serious barrier to the comparison of pulsational and evolutionary masses for RRd variables. By contrast to their affect upon pulsational masses, the Livermore OPAL opacities appear to have only a small effect upon the masses of RR Lyrae stars as determined from evolutionary tracks of horizontal branch stars (Yi et al. 1993).

5.3.3   *Other kinds of mixed mode behavior*
There are a few reports in the literature of RR Lyrae stars which apparently show multiperiodic behavior which is of neither the Blazhko nor the RRd variety. AC Andromedae and ST Canum Venaticorum in particular have been subjected to detailed analysis.

AC And is a metal-rich variable star which has a primary pulsation period of 0.711 days, which would place it in the period realm appropriate to RR Lyrae stars. Its lightcurve does not, however, repeat exactly with the 0.711 day period, and analysis of an extensive series of Konkoly Observatory observations revealed the existence of additional periods of 0.525 and 0.421 days (Fitch and Szeidl 1976). The amplitude of the 0.711 day period was strongest, closely followed by the 0.525 day period, with the 0.421 day period being somewhat weaker. Fitch and Szeidl interpreted these periodicities as those of the fundamental, first overtone, and second overtone radial modes, respectively. The pulsational mass they calculated, 3.1 $M_\odot$, was far in excess of that expected of an RR Lyrae star, and in excess of the 0.61 $M_\odot$ value Stellingwerf (1975) had earlier calculated on the basis of just the two longer periods. Cox et al. (1978) obtained a similarly large value for the mass of AC And. The interpretation of the multiple periodicities of AC And is still uncertain, but there remains a possibility that, despite its primary period, AC And is not an RR Lyrae variable. It may be related to the Population I Cepheids or to variable stars of the δ Scuti type (Kurtz 1988; Petersen 1990b).

ST CVn is another unusual star. It appears to be an RRc star with a primary period of 0.329 days. Peniche et al. (1989) analyzed visual, photographic, and photoelectric observations of ST CVn and found evidence for additional periodicities of 0.161 and 1.324 days. The longer period is too long to be the fundamental mode and the shorter period, nearly half the 0.329 day period, is too short to be the second overtone period, according to pulsation models of RR Lyrae variables. Additional observations of ST CVn would be of value.

5.4 Future directions

New theoretical results may refine and extend our understanding of RR Lyrae period changes, but significant new observational evidence for period changes can only be accumulated at a frustratingly slow rate. Nevertheless, it is important to continue the observations of field and cluster RR Lyrae variables so that future astronomers will have the data to compare theoretical and observed rates of evolution for horizontal branch stars.

While new observational results for RRd and Blazhko effect stars can be accumulated at a faster rate, it is clear that advances in theory are needed to improve our understanding of multiperiodic RR Lyraes. For the Blazhko effect stars, the basic phenomenon is still poorly understood, and there is a need for theoretical models which yield testable predictions. For the RRd variables, the origin of the double mode pulsation and the different proportions of RRd stars in different systems remain without adequate explanation. Work remains, too, before the RRd stars can be made to give reliable masses for comparison with those obtained by stellar evolution theory.

6

RR Lyrae stars beyond the Milky Way

The search for RR Lyrae stars in galaxies beyond our own has been pressed for two reasons: first, because RR Lyrae stars can be used to determine the distances to systems in which they are found; and second, because the frequency and distribution of RR Lyrae stars in a system inform us about its old stellar population. However, the faintness of RR Lyraes relative to the classical Cepheids retarded the discovery of extragalactic RR Lyrae stars. Some 60 years passed between Hubble's identification of Cepheids in the Andromeda galaxy and Pritchet and van den Bergh's observations of its RR Lyrae stars. Nonetheless, a considerable body of observation has now accumulated about RR Lyrae stars in systems beyond the Milky Way and, with the application of charge-coupled device (CCD) detectors on large telescopes, the field is now an active one. It is, however, a curiosity that, as will be discussed below, RR Lyrae stars perhaps made their most important contribution to fixing the extragalactic distance scale by being too faint for Walter Baade to detect on photographs taken with the Palomar 5-m telescope. In this chapter we shall review what is known of RR Lyrae stars beyond the Milky Way, beginning with its relatively near companions, the Magellanic Clouds and dwarf spheroidal systems, and concluding with the more distant Local Group members.

6.1 The Magellanic Clouds

The Large and Small Magellanic Clouds (LMC and SMC), distant some 50 and 60 kpc, respectively, are the nearest systems to the Milky Way displaying both old stars and ongoing star formation. Photographic plates taken at the Arequipa, Peru, station of Harvard College Observatory showed that both of these irregular stellar systems contain many Cepheid variable stars. It was through an investigation of the Cepheids in the SMC that Henrietta Leavitt (1908, Pickering 1912) discovered that Cepheids obey a period–luminosity relation. Observations of the Magellanic Clouds therefore hold the potential of tying together the luminosities of the RR Lyrae stars and the classical Cepheids. However, the story of RR Lyrae stars in the Magellanic Clouds has a false beginning. Shapley (1922) reported finding 13 RR Lyrae variables in the Small Magellanic Cloud. These variables are far brighter than the actual SMC RR Lyraes and Payne-Gaposchkin and Gaposchkin (1966) were able to show that Shapley's periods for these stars were spurious (Fernie 1969). Dartayet and Dessy (1952) also reported the discovery of SMC variables with periods under one day, which they assumed to be RR Lyrae variables. However, these stars, of apparent magnitude 17, are about 2 magnitudes brighter than actual SMC RR Lyraes and may be variables of the anomalous Cepheid type (§6.3) or very short period classical Cepheids (Smith et al. 1992).

The actual discovery of RR Lyrae stars in the Magellanic Clouds came three decades after Shapley's spurious discovery. After the Second World War, the Radcliffe 1.9-m reflector went into operation in South Africa. It was at that time the largest southern hemisphere telescope and one of the first tasks to which it was put was a search for RR Lyrae stars in the Magellanic Clouds. The outcome of this successful search was the discovery of RR Lyrae stars near the 19th magnitude in both the Large and Small Clouds (Thackeray 1951; Thackeray and Wesselink 1953).

As Thackeray (1974) later wrote, this discovery had a twofold impact: 'Firstly, it proved that both Clouds contain old populations as well as the conspicuous young Population I...' As late as 1950, Baade (1951) had pointed to the absence of RR Lyrae variables in the Magellanic Clouds as evidence of the Clouds' essentially Population I nature. 'Secondly, the fact that the variables appeared at the 19th magnitude instead of the expected 17.5 immediately confirmed Baade's 1952 revision of the distance scale'. The faintness of the RR Lyrae stars in the SMC was in fact announced by Thackeray at the same 1952 Rome meeting of the IAU at which Baade reported his failure to find RR Lyraes in the Andromeda Galaxy with the Palomar 5-m telescope (§6.4). Subsequent studies of Magellanic Cloud RR Lyraes have continued these two lines of investigation, contributing to our understanding of the structures and histories of the Clouds while helping to better fix the distances to these comparatively nearby extragalactic systems (or, alternatively, fixing the absolute magnitudes of the RR Lyraes).

6.1.1 RR Lyraes in Magellanic Cloud clusters

Magellanic Cloud RR Lyraes were first identified in and near NGC 121, an old star cluster in the Small Magellanic Cloud. RR Lyraes have been found subsequently in several star clusters of the LMC, but no other SMC star cluster has been observed to contain such variables. Table 6.1, which is an updated and enlarged version of table 1 in Graham and Nemec (1984), lists the Magellanic Cloud clusters in which RR Lyraes have been discovered and summarizes some of the results to date. $\langle P_{ab} \rangle$ is, as usual, the mean period of the RRab stars in each cluster. The RR Lyrae stars in Magellanic Cloud clusters are relatively faint and often lie in crowded fields. Systematic errors of 0.1 mag in photographic photometry of such stars are not uncommon. In giving mean magnitudes in table 6.1, I have relied upon the CCD photometry of Walker and collaborators when such was available.

There is some evidence for a division of the Magellanic Cloud clusters into the two Oosterhoff groups, but the distinction may be less sharp than in the Galaxy. Several clusters have $\langle P_{ab} \rangle = 0.57$–$0.59$ days, long for Oosterhoff I but short for Oosterhoff II. In addition, whereas the mean period of RRab variables in NGC 2257 is nearer the canonical value for Oosterhoff I than Oosterhoff II clusters, its relatively large number of RRc variables is characteristic of an Oosterhoff II cluster. It therefore cannot be easily assigned to either Oosterhoff group.

The population of bright star clusters of the Magellanic Clouds differs strikingly from that of the Galaxy in at least one respect. Unlike the Galaxy, the Magellanic Clouds contain populous, globular-like, star clusters which are young as well as old. Searle, Wilkinson, and Bagnuolo (1980) classified these populous clusters on the basis

Table 6.1. *Magellanic Cloud clusters with RR Lyrae stars*

Cluster	No. RR	$\langle B \rangle$	$\langle V \rangle$	$\langle P_{ab} \rangle$	Reference
SMC					
NGC 121	4	19.93	19.59	0.55	1, 2, 3, 4
LMC					
NGC 1466	43	19.1:	19.05	0.59	5, 6, 20, 21
NGC 1786	11	19.57	19.27	0.66	7, 8
NGC 1835	23	19.4:	19.37	0.59	9, 20, 21
NGC 1841	22	19.77	19.31	0.68	10, 11
NGC 2210	40:	19.4	19.12	0.64	12, 13, 14
NGC 2257	41	19.3	19.03	0.58	15, 16, 17
Reticulum	32	19.38	19.07	0.55	18, 19, 20

References
1. Thackeray 1958
2. Tift 1963
3. Graham 1975
4. Walker and Mack 1988a
5. Wesselink 1971
6. Norris 1973
7. Graham 1985
8. Walker and Mack 1988b
9. Graham and Ruiz 1974
10. Kinman, Stryker, and Hesser 1976
11. Walker 1990
12. Graham and Nemec 1984
13. Walker 1985
14. Hazen and Nemec 1992
15. Alexander 1960
16. Nemec, Hesser, and Ugarte 1985
17. Walker 1989b
18. Demers and Kunkel 1976
19. Gratton and Ortolani 1987
20. Walker 1992a
21. Walker 1992b

of photometry of their integrated light in the Gunn–Thuan uvgr photometric system. This classification system has turned out to be generally useful as a shorthand characterization of cluster ages. The youngest and bluest star clusters are classified as type I while the oldest and reddest clusters, which appear to be similar in age and metallicity to the globular clusters of the galactic halo, are classified as type VII.

All of the clusters in which RR Lyraes have been found are classified as type VII on this photometric system, i.e., they are all very old. However, not all type VII clusters appear to contain RR Lyraes. Graham and Nemec (1984) searched in vain for RR Lyraes in the type VII clusters NGC 339, NGC 2019, and Hodge 11 [The type VII classification for NGC 339 has, however, subsequently been queried by Olszewski (1988) and Smith, Searle, and Manduca (1988)]. No RR Lyrae stars have been discovered in clusters which Searle, Wilkinson, and Bagnuolo classified as type VI, their second oldest category (Graham and Nemec 1984, Walker 1989a). That some old Magellanic Cloud clusters lack RR Lyraes should not surprise us, considering the wide range in frequency of RR Lyrae stars among the globular clusters of the Galaxy. More important is that there appears to be a lower cutoff in age for those Magellanic Cloud clusters which contain RR Lyrae stars, a cutoff which may tell us something about which stars can evolve to become RR Lyraes.

Stellar evolution theory tells us that RR Lyrae stars are low-mass stars. Nevertheless, because whether a red giant evolves into the instability strip during its horizontal branch lifetime depends upon poorly known quantities such as the amount of mass loss it sustains prior to the horizontal branch stage, the age of the youngest RR Lyrae stars cannot be accurately predicted by theory. The point is of interest because the presence of RR Lyrae stars in a system is taken as indicating the existence of a very old stellar population. It would be useful if this statement could be quantified by setting a minimum age to the population represented by the RR Lyrae stars. The populous Magellanic Cloud clusters, with their wide range in age, allow one to set such a limit.

What is the age of the youngest Magellanic Cloud cluster which contains RR Lyrae stars? Stryker, Da Costa, and Mould (1985) determined the age of NGC 121 to be 12 ± 2 Gyr, for which determination they assumed $M_v = +0.6$ for its RR Lyrae stars. This would make NGC 121 comparable in age, but perhaps slightly older than, Lindsay 1, to which Olszewski, Schommer, and Aaronson (1987) assigned an age of 10 Gyr. Both NGC 121 and Lindsay 1 are believed to have [Fe/H] values near -1.4. Olszewski, Schommer, and Aaronson argued that, because NGC 121 contains RR Lyrae stars while none have yet been found in Lindsay 1 (Gascoigne 1966, Walker 1989a), an age of about 10–12 Gyr may be the youngest age at which clusters of [Fe/H] $= -1.4$ can make RR Lyraes. If this can be taken as a general guide, the existence of RR Lyrae stars does signify the occurrence of a very old stellar population, but that population perhaps need not be quite as old as the globular clusters of the Galaxy. A note of caution must be sounded at this point, however: from the absence of RR Lyraes in Magellanic Cloud clusters which are younger than 12 Gyr of age it may not necessarily follow that younger RR Lyraes cannot be produced elsewhere, in circumstances different from those found in the Clouds.

Of all the Magellanic Cloud clusters, NGC 2257 is the one for which the most detailed studies have been made of large numbers of RR Lyrae stars. Its RR Lyrae stars have been studied photographically by Alexander (1960) and Nemec, Hesser, and Ugarte (1985) and with CCD photometry by Walker (1989b). Of the RR Lyrae stars in Magellanic Cloud clusters, only for those in NGC 2257 is there already a sufficient observational record to allow their period change behavior to be studied (see Table 5.1).

6.1.2 Field RR Lyraes in the Magellanic Clouds
Several studies have been made of the general field RR Lyrae population in the Magellanic Clouds but no comprehensive survey of field RR Lyrae stars in the Clouds has been completed and, in general, much remains to be done along these lines. The survey of RR Lyrae stars in the LMC by Reid and Freedman (1990), in progress as I write, may help to improve this situation. A number of important points can, however, already be addressed, including the question of whether the field RR Lyraes in the Magellanic Clouds resemble their counterparts in the Milky Way.

The Small Magellanic Cloud
In the SMC, there have been two surveys of field RR Lyraes: Graham (1975) investigated RR Lyrae stars in an outlying field near NGC 121, while Smith et al. (1992) studied RR Lyrae stars in a field near the Northeast Arm of the SMC. Both studies are based upon similar plate material obtained at the 1.5-m telescope of Cerro Tololo Interamerican Observatory (Chile), and each study investigated variables in a 1×1.3 degree region. Results of these investigations are summarized in table 6.2.

Graham identified 76 field RR Lyraes near NGC 121. The period histogram for

Table 6.2. *Field RR Lyraes in the Magellanic Clouds*

Field	Type	No.	Mean Period	$\langle B \rangle$	$\langle V \rangle$
LMC N1783 field	ab	60	0.564 d	19.61	19.2
	c	8	0.328	19.56	19.3
LMC N2257 field	ab	13	0.564	19.45	19.18
	c	11	0.363	19.5	—
LMC N2210 field	ab	40	0.576	19.61	19.35
	c	12	0.345	19.46	19.32
SMC N121 field	ab	55	0.589	19.95	19.6
	c	14	0.378	19.92	19.6
SMC NE Arm	ab	17	0.57	19.91	—
	c	5	0.36	19.94	—

these stars is shown in figure 6.1, while their period–amplitude diagram is shown in figure 6.2. One aspect of figures 6.1 and 6.2 offers a striking contrast with the situation in the solar neighborhood: there is a conspicuous absence of RRab stars with periods less than 0.45 day. Such stars are common in the solar neighborhood (Preston 1959), making up perhaps 20 percent of the RR Lyraes brighter than 13th magnitude. Preston's ΔS study showed these short period RRab variables to be relatively metal-rich stars belonging to the thick disk population of the Galaxy (see chapter 4). The absence of similar variables in the NGC 121 region suggests that the SMC may contain no similar old, metal-rich, stellar population. This is consistent with independent lines of evidence which suggest that the SMC has not reached the same stage of chemical enrichment as has the galactic disk.

The solid line in figure 6.2 is the period–amplitude relation for RRab stars in the globular cluster M3 (NGC 5272). At a given period, SMC RR Lyraes tend to have smaller B amplitudes than their M3 counterparts, though Graham warns that the amplitudes of some of the SMC RR Lyraes are uncertain. According to the Sandage period shift effect (§3.5), this indicates that the SMC variables are, on average, slightly more metal-rich than M3, i.e., more metal-rich than about [Fe/H] = −1.6. The scatter in the period-amplitude diagram indicates, however, that there may be a significant star-to-star range in [Fe/H]. Butler, Demarque, and Smith (1982) measured ΔS metal abundances for three RR Lyrae stars in the Graham field, obtaining $\langle$[Fe/H]$\rangle$ = −1.8 ± 0.2 on Butler's metallicity scale.

The field around NGC 121 is evidently one in which there has been little recent star formation. Stryker, Da Costa, and Mould (1985) deduced from color–magnitude data that the NGC 121 region is dominated by an old population, but could not exclude the existence of a small component of younger stars. Suntzeff et al. (1986) found the average metallicity of K-giants in the NGC 121 region to be [Fe/H] = −1.56, very similar to that expected for the RR Lyrae stars on the basis of the period–amplitude diagram. The K-giants thus appear to represent an old population similar to that represented by the RR Lyrae stars. Both the K-giants and the RR Lyraes have probably evolved from very similar main sequence progenitors.

In the Northeast Arm region, Smith et al. (1992) identified 42 probable field RR Lyrae stars, but determined periods for only 22 of these. These 22, however, seem

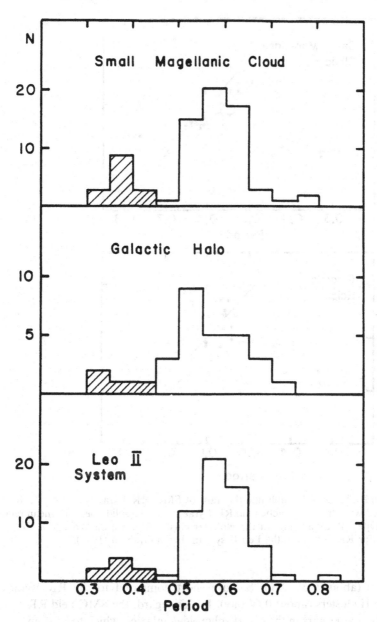

Figure 6.1 The period histogram for field RR Lyrae stars in the SMC near NGC 121. For comparison the period histograms for RR Lyrae stars in the galactic halo toward the North Galactic Pole (Kinman et al. 1966) and in the Leo II dwarf spheroidal system are shown. Hatched areas indicate RRc variables. From Graham (1975).

generally to occupy the same area of the period–amplitude diagram as the RR Lyrae stars in the field around NGC 121. With one possible exception, the period–amplitude diagram again provided no evidence for the existence of a metal-rich RR Lyrae population.

The mean periods of RRab stars in the two SMC fields (table 6.2) appear to be unusually long compared to the mean periods of RR Lyraes in Oosterhoff type I

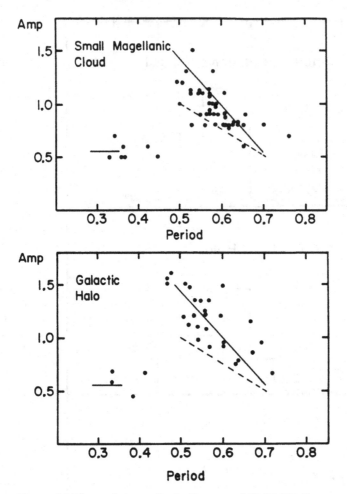

Figure 6.2 The period–amplitude diagram of SMC RR Lyrae stars near NGC 121, compared with that for galactic halo RR Lyrae stars. The solid line is the mean relation for ab-type RR Lyrae stars in the globular cluster M3; the dashed line is the mean relation for RRab stars in the Leo II system. From Graham (1975).

globular clusters (about 0.55 day), but are short compared to the RR Lyraes in Oosterhoff type II clusters (about 0.64 day). In this regard, the SMC field RR Lyraes are similar to RR Lyrae stars in the dwarf spheroidal galaxies, which to an even greater degree tend to have values of $\langle P_{ab} \rangle$ intermediate between those of the Oosterhoff I and II globular clusters.

When the RR Lyrae surveys of Graham (1975) and Smith et al. (1992) are corrected for incompleteness, it is found that the NGC 121 field and the Northeast Arm field contain comparable numbers of field RR Lyraes: about 90–95 for the NGC 121 field and 81–86 for the Northeast Arm field. The NGC 121 field is more distant from the main body of the SMC than the Northeast Arm field, so that this result implies that RR Lyrae stars in the SMC are not strongly concentrated to the main body of that system. To some degree, we may take the distribution of RR Lyrae stars in the SMC as representative of the distribution of the old population as a whole. The main body of the SMC may therefore be a concentration mainly of the intermediate and young stars of that system, with the oldest populations having a more extended distribution.

Graham found the SMC RR Lyraes to peak sharply at $\langle B \rangle = 20.0$. He estimated that the true dispersion in apparent magnitude about the peak is unlikely to be greater than 0.1 mag. Thus, near the outlying cluster NGC 121 there is little sign of an extensive front to back distance spread as has been suggested by some observations of SMC Cepheids (Mathewson et al. 1986).

Though only two SMC fields have been investigated, and data for the Northeast Arm field are still incomplete, the similarity of the results for the two fields is notable. They have about the same surface density of RR Lyraes, the mean periods of RRab and RRc stars in the two fields are similar, and RR Lyraes in the two fields occupy the same region in the period–amplitude diagram. Overall, there is a likeness between these two fields which argues for a homogeneity of RR Lyrae properties in different regions of the SMC.

The Large Magellanic Cloud

In the LMC substantial numbers of field RR Lyraes have been observed in three areas: around NGC 1783, around NGC 2210, and around NGC 2257. All three of these areas are relatively far from the HI center of mass of the LMC, with projected distances of about 3.4 kpc, 3.9 kpc, and 7.4 kpc, respectively (Kinman et al. 1991). Graham (1977) identified and determined periods for 68 field RR Lyrae stars in a region around NGC 1783, a region populated by both young and old stars. Hazen and Nemec (1992) presented data on 52 field RR Lyrae stars in the vicinity of the LMC cluster NGC 2210. Finally, Nemec, Hesser, and Ugarte (1985) discovered 47 variable stars in the field around NGC 2257. They determined periods and light curves for 24 field RR Lyraes in this region and Walker (1989b) obtained B, V CCD photometry for nine of these stars. Nemec, Hesser, and Ugarte suspected two of the variables in this field of showing mixed mode behavior, but Walker (1989b) found that both of these stars were actually RRab variables.

The distribution over period of the RR Lyraes in the three LMC fields is shown in figure 6.3, while the period–amplitude diagrams for the three fields are given in figure 6.4. Summaries of the periods and apparent magnitudes of RR Lyraes in the LMC fields are tabulated in table 6.2. For the mean periods of RRab and RRc stars in the NGC 2257 field, I have adopted the results of Nemec et al. (1985). However, Walker (1989b, Kinman et al. 1991) has slightly corrected the Nemec et al. photometric scale, and I give his value for the mean magnitude of the RRab stars. A similar correction presumably applies to the Nemec et al. RRc magnitudes, but as this is not given explicitly by Walker, only an approximate value for this magnitude is listed in table 6.2.

Hazen and Nemec (1992) drew attention to certain similarities and differences among the RR Lyrae populations of the LMC fields. The mean periods for the RRab and RRc stars are similar in the three fields, and are all slightly high compared to the averages for Oosterhoff I globular clusters. The mean periods tend, however, to be shorter than those found for RR Lyrae stars in dwarf spheroidal systems. Whereas the RR Lyrae stars in the two SMC fields appear to represent similar populations, there may be important differences among the three LMC fields. The number density of RR Lyrae stars decreases in going from the NGC 1783 field to the NGC 2210 field to the

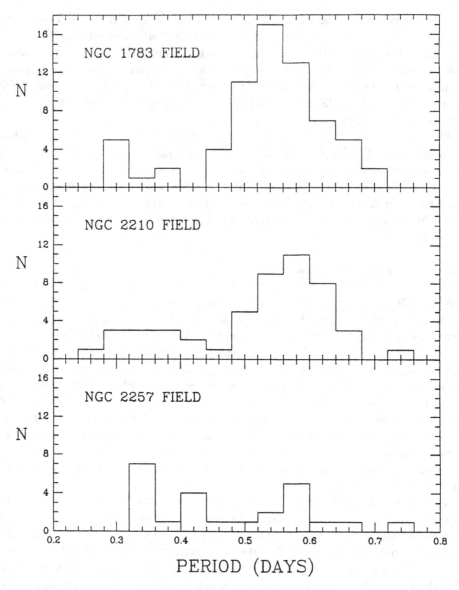

Figure 6.3 Period histogram diagrams are shown for field RR Lyraes near NGC 1783, NGC 2210, and NGC 2257. From Hazen and Nemec (1992).

NGC 2257 field, and so does the ratio of RRab to RRc stars. Provided that it cannot be explained as a bias in the discovery statistics, this trend in the N(RRab)/N(RRc) ratio may reflect a trend in metal abundance. Recall that, in the Galaxy, the N(RRab)/N(RRc) ratio is smaller for the very metal-poor Oosterhoff II globular clusters than for the less metal deficient Oosterhoff I globular clusters. Thus, metallicity may decrease in going from the NGC 1783 field to the NGC 2257 field.

Hazen and Nemec found additional support for such a metallicity trend in the period–amplitude diagrams for the three fields. Comparing these diagrams with those of RR Lyrae stars in the globular clusters M3 and M15, they deduced a mean [Fe/H] of

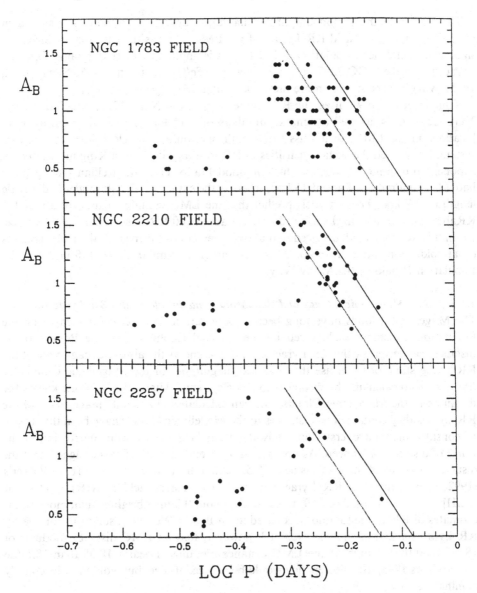

Figure 6.4 Period–amplitude diagrams are shown for field RR Lyrae stars near NGC 1783, NGC 2210, and NGC 2257. The solid line indicates the mean period–amplitude relation for RRab variables in M15. The dotted line indicates the mean period–amplitude relation for RRab variables in M3. From Hazen and Nemec (1992).

about −1.3 for RR Lyraes in the NGC 1783 field and −1.8 for RR Lyraes in the NGC 2210 field. The smaller number of stars made their conclusions for the NGC 2257 field less certain, but the mean metallicity suggested by its period–amplitude relation does not appear to differ significantly from that derived for RR Lyraes in the NGC 2210 field. Butler et al. (1982c) obtained ΔS metallicity measures for seven RR Lyrae stars in the NGC 1783 field, obtaining [Fe/H] = −1.4 ± 0.2, a result in agreement with that deduced from the period–amplitude diagram.

The scatter in the period–amplitude diagrams suggests that there is a spread in metallicity among the field RR Lyraes of the LMC. As was the case for the SMC, no significant metal-rich population of RR Lyraes is implied by these diagrams. One RR Lyrae star in the NGC 2257 field may have an [Fe/H] as high as -0.5 (Nemec et al. 1985), while others may be more metal-poor than M15 ([Fe/H] < -2.15).

Kinman et al. (1991) used the RR Lyrae stars in the NGC 1783, NGC 2210, and NGC 2257 fields, as well as smaller numbers of field RR Lyraes observed at other locations in the LMC, to discuss the overall distribution of RR Lyrae stars in that system. They fitted the surface densities of RRab stars with both a King model and an exponential disk model. The best fit King model (a lowered Maxwellian – King 1966) had a limiting radius of about 15 kpc, whereas the exponential disk model had a scale length of 2.7 kpc. These models predict that the LMC contains a total of about 10^4 RRab stars, with a central surface density of about 200 RRab stars per square degree. From this, Kinman et al. deduced that about 2 percent of the mass of the LMC consists of an old, Population II halo. This percentage is similar to that found for the Population II halo of the Milky Way.

6.1.3   *The Magellanic Clouds and the absolute magnitude of the RR Lyrae stars*

The Magellanic Clouds have long been seen as an ideal location for comparing the Population I distance scale given by the classical Cepheids and the Population II distance scale given by the RR Lyrae stars. If one knows the absolute magnitude of the RR Lyrae stars, one can use the observed magnitudes of the Magellanic Cloud RR Lyrae stars to calculate the distances to those systems. Alternatively, if one knows the distances to the Magellanic Clouds, one can calculate the absolute magnitudes of the RR Lyraes they contain. The distances to the Magellanic Clouds have been the subject of considerable controversy, but lately there has been a significant narrowing of the range of distances advanced. As noted in §2.9, several lines of evidence suggest that the distance modulus of the LMC is near 18.5. With a distance modulus of 18.5, Walker's (1992c) observations of RR Lyrae stars in LMC clusters yield $M_v(RR) = +0.44$ at [Fe/H] $= -1.9$. As noted in §2.9, this result is about 0.3 mag brighter than most recent estimates of RR Lyrae luminosities based upon Baade–Wesselink solutions for nearby RR Lyrae stars. It cannot be excluded, however, that an LMC distance modulus of 18.5 is slightly too high. If the LMC distance modulus is nearer 18.38 than 18.5 (de Vaucouleurs 1993), the discrepancy would be reduced in size, but would not be entirely eliminated.

Walker (1991) presented a preliminary consideration of the differences in the distances to the Magellanic Clouds based upon CCD photometry of cluster RR Lyrae stars. He concluded that, after correcting for absorption and geometric projection, RR Lyrae stars in the SMC cluster NGC 121 were 0.52 ± 0.08 magnitudes fainter in V than RR Lyraes in LMC clusters. Walker pointed out that not all of this difference may be attributable to differing distances of the LMC and SMC, because the RR Lyrae stars in NGC 121 may differ in absolute magnitude from those in the LMC clusters. NGC 121 is somewhat more metal-rich than the LMC clusters, and therefore any correlation between metallicity and absolute magnitude among the RR Lyrae stars will enter into the comparison. The difference in apparent magnitude between the field RR Lyraes of the LMC and SMC (table 6.2) appears to be about 0.4 ± 0.1. For these field stars, the difference in mean metallicity, as deduced from the period–amplitude relations, appears to be at most only a few tenths in [Fe/H]. Because interstellar extinction appears to be small in these fields, most of this difference in apparent magnitude can probably be attributed to differing distances to the LMC and SMC. Thus, if the distance modulus of the LMC is about 18.5, that of the SMC is probably about 18.9.

6.2 RR Lyrae Stars in dwarf spheroidal galaxies

Including the recently discovered Sextans system, eight dwarf spheroidal galaxies appear to be associated with the Milky Way. All are more distant than the Magellanic Clouds with the most distant of them, Leo I and Leo II, being about 250 kpc away. They are systems of low surface brightness and low metallicity and, unlike stars within a globular cluster, stars within a dwarf spheroidal system exhibit a range in [Fe/H] and, in some cases, a range in age as well. Zinn (1985b) and Da Costa (1988) have reviewed the properties of these systems.

Shortly after Shapley reported in 1938 the discovery of Sculptor and Fornax, the first two dwarf spheroidals to be found, Baade and Hubble (1939) observed them with the Mt. Wilson 2.5-m telescope. Both systems lie near declination −34°, which complicated observations from as northerly a location as Mt. Wilson. Nonetheless, Baade and Hubble were able to identify 40 variable stars in the Sculptor system. Although they were unable to determine periods for these stars, they correctly identified the bulk of them as being cluster-type variables, that is, RR Lyrae stars. The recognition of RR Lyrae variables in a dwarf spheroidal system thus preceded their discovery in the Magellanic Clouds by some dozen years.

RR Lyrae stars have now been identified in six dwarf spheroidal systems, Fornax and Leo I being the exceptions. However, color–magnitude diagrams of Fornax clusters and field stars suggest that RR Lyraes probably exist in that system, too (Buonanno et al. 1985). A detailed study of the RR Lyrae population in the Sextans system has not yet been published, but Weller, Mateo, and Krzeminski (1991) reported the discovery of about 40 RR Lyrae stars in the Sextans system. The color–magnitude diagram of Sextans by Mateo et al. (1991) shows a well-populated horizontal branch in the vicinity of the instability strip. Mateo et al. concluded that the Sextans system may contain as many as 300 RR Lyraes.

Leo I may be the sole dwarf spheroidal system without RR Lyrae stars. The proximity of Regulus hindered photographic observations of Leo I, but Hodge and Wright (1978) detected some variable stars which they believed were probably RR Lyrae stars caught near maximum light. However, the deep color–magnitude diagram of Leo I by Lee et al. (1993) did not show the existance of any stellar population in Leo I older than about 3 Gyr. In particular, they saw no evidence of an old horizontal branch population in their color–magnitude diagram of 16 000 Leo I stars.

Table 6.3 summarizes results obtained for RR Lyrae variables in the dwarf spheroidal systems. Values of ⟨[Fe/H]⟩ and M_v for the dwarf spheroidal systems are adopted, with some rounding, from Zinn (1985b), except for that of Sextans, which is taken from Mateo et al. (1991). In the column headed 'No.' are the number of variables reported discovered in each dwarf spheroidal system; most of these are believed to be RR Lyraes. Tabulated under the columns headed 'ab' and 'c' are the number recognized as ab or c type RR Lyrae stars and having reliable periods.

A large proportion of the dwarf spheroidal systems have ⟨P_{ab}⟩ values intermediate between the canonical values for Oosterhoff I and Oosterhoff II globular clusters,

Table 6.3. *RR Lyrae stars within dwarf spheroidals*

System	$\langle[Fe/H]\rangle$	M_v	No.	ab	c	$\langle P_{ab}\rangle$	Reference
Draco	−2.2	−8.5	261	116	16	0.61	Baade and Swope 1961
							Nemec 1985b
Ursa Minor	−2.2	−8.8	92	47	35	0.64	van Agt 1967
							Nemec et al. 1988
Carina	−1.9	−9.2	58	48	9	0.62	Saha et al. 1986
Leo II	−2.0	−10.2	196	64	6	0.59	Swope 1967
							van Agt 1973
Sculptor	−1.8	−11.6	380:	51	9	0.60	van Agt 1978
							Goldsmith 1993
Sextans	−1.5	−9.5	40	—	—	—	Weller et al. 1991

about 0.55 and 0.64 days (van Agt 1973). Even though it is now realized that the clusters within each Oosterhoff group exhibit a range in $\langle P_{ab}\rangle$, the number of dwarf spheroidal systems with $\langle P_{ab}\rangle$ between 0.57 and 0.62 days is remarkable. Clearly, the period distributions of RRab stars in the dwarf spheroidal systems differ from those of their globular cluster counterparts. The differences in the period histograms for RR Lyraes in dwarf spheroidal systems and in archetypal Oosterhoff I and II globular clusters are illustrated in figure 6.5.

The Draco and Ursa Minor RR Lyraes have been the most extensively studied. Nemec (1985b), reanalyzing the photometry of Baade and Swope (1961), identified ten double mode RR Lyrae stars in Draco. These variables show a mixture of fundamental and first overtone pulsation modes. Nine of these stars appear to have pulsational masses similar to the mixed mode stars in the Oosterhoff II cluster M15, while one has a pulsational mass similar to the double mode pulsators in the Oosterhoff I cluster M3 (§5.3). In contrast, Nemec, Wehlau, and Oliveira (1988) found no double mode RR Lyraes in their investigation of variables in the Ursa Minor dwarf galaxy.

The location of RR Lyraes in the period–amplitude and period–rise time diagrams has been investigated for Draco (Nemec 1985b) and Ursa Minor (Nemec et al. 1988). In Draco Nemec discovered that the most luminous RR Lyrae stars showed, on average, the largest period shifts relative to the fiducial line defined by the RR Lyrae stars in M3. It follows from the discussion of the period shift effect in §3.5 that a range in period shift may imply a spread in [Fe/H] among the Draco RR Lyrae stars. The implied [Fe/H] spread is at least as great as the difference in metallicity between M3 ([Fe/H] = −1.7) and M15 ([Fe/H] = −2.2). The existence of such a star-to-star spread in metallicity among the RR Lyrae stars is consistent with observations of Draco's red giant stars, which also show a star-to-star range in metallicity (Zinn 1985b and references therein). Some caution about this interpretation arises, however, from the circumstance that evolution away from the ZAHB can also affect the magnitude of the period shift. Nemec et al. (1988) found that in UMi the long period RRab variables have more asymmetric light curves than the shorter period RRab variables. This is contrary to the case in the globular clusters of the Galaxy and in Draco but, as Nemec et al. note, there is enough scatter in their results to make this conclusion uncertain.

Deep color–magnitude diagrams for seven dwarf spheroidals have recently allowed

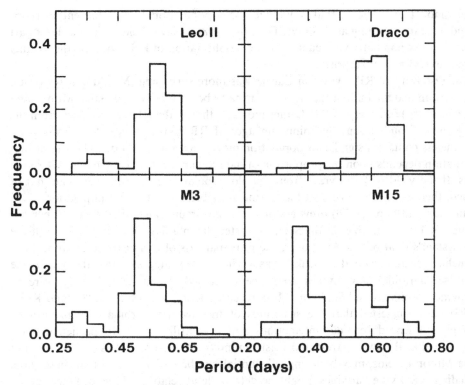

Figure 6.5. Period histograms for RR Lyrae stars in Draco and Leo II are compared with those for RR Lyrae stars in M3 (Oosterhoff I) and M15 (Oosterhoff II). After Kukarkin (1975).

the ages of stars in those systems to be determined. As mentioned above, the vast majority, and perhaps all, of the stars in the Leo I system appear to be no older than about 3 Gyr. Olszewski and Aaronson (1985) found the color–magnitude diagram of Ursa Minor to be well fitted by an isochrone for age 16 ± 2 Gyr, an age comparable to that of a globular cluster of the Galaxy. There is little evidence for a younger population in UMi, though some blue stragglers are present. Likewise, Mateo et al. (1991) found no evidence for a significant population in Sextans younger than 12 Gyr, and possibly none younger than 16 Gyr. Results for Draco (Stetson, VandenBerg, and McClure 1985; Carney and Seitzer 1986) also indicate a predominantly old population, though Carney and Seitzer suggested that a sizeable number of stars in Draco may be a few Gyr younger than those in a typical globular cluster. Da Costa (1984) has argued that the bulk of the stars in Sculptor are 2–3 Gyr younger than those in globular clusters of the Galaxy, but at 12–13 Gyr, they are still quite old. As all of these systems contain a large proportion of old stars, the existence of RR Lyrae stars in them is not surprising. However, the cases of Carina and Fornax are more complicated.

Fornax contains five, possibly six, globular clusters. In all observations to date, these globular clusters appear similar to those of the galactic halo (Buonanno et al. 1985), though the main sequence turnoffs of these clusters have not yet been observed. This suggests the presence of a very old stellar population in Fornax, but the large number of fairly bright carbon stars in Fornax and the deep color–magnitude diagrams of

Buonanno et al. indicate that a younger field population is also present, perhaps including stars as young as 2–3 Gyr (Da Costa 1988). Nevertheless, since at least part of the Fornax population appears to be old, confirmation of RR Lyraes in the Fornax system would not be surprising.

The discovery of RR Lyraes in Carina was more surprising. Mould and Aaronson (1983) estimated the bulk of the stars in Carina to be only 7 Gyr old. Saha, Monet, and Seitzer's (1986) discovery of RR Lyraes indicates that, if the inferences drawn from the Magellanic Clouds about the minimum ages of RR Lyrae stars hold, Carina must nonetheless contain a significant population older than about 10 Gyr. The size of that population depends upon the proportion of old red giants which evolve into RR Lyrae stars. If the yield of RR Lyraes from the old population in Carina were like that of Draco, then about 15 percent of the Carina stars belong to this old population.

Innanen and Papp (1979) drew attention to the circumstance that about 10 percent of the RR Lyrae variables in the Sculptor system lie in a flattened distribution outside that system's tidal radius. Moreover, the principal axis of this distribution appears to be inclined relative to that of Sculptor as a whole. They suggested that the occurrence of these extratidal RR Lyraes could be explained dynamically if they were in retrograde orbits around Sculptor. More recently, Kuhn and Miller (1989) and Kuhn (1993) have suggested that the structures of the dwarf spheroidal systems can be strongly influenced by tidal interactions with the Milky Way. In particular, they proposed that the two nearest and least luminous dwarf spheroidals, the Draco and Ursa Minor systems, may be losing substantial numbers of stars and that these stars, including RR Lyrae variables, might be detectable at relatively large distances along the major axes of these systems.

6.3   Anomalous Cepheids

Though they are not RR Lyrae stars, and therefore in strict construction are beyond the scope of this book, there are unusual Cepheid variable stars in the dwarf spheroidal systems which deserve brief mention because their pulsation periods are similar to those of the RR Lyrae stars. These stars are known as anomalous Cepheids because they have been believed to obey neither the type I nor the type II Cepheid period–luminosity relations. The periods of the anomalous Cepheids range between 0.4 and 2 days, thus overlapping with the RR Lyrae stars at the short period end of the range, but they are 0.5–2 magnitudes brighter in V than are the RR Lyrae stars. Anomalous Cepheids have been identified in six of the Milky Way's companion dwarf spheroidal systems and it seems likely that they occur in a seventh, Carina, as well. In Carina, Saha, Monet, and Seitzer (1986) identified eight variables which they identified as foreground ab-type RR Lyrae stars. These suspected foreground variables are 0.5–1.5 magnitudes brighter than the Carina RR Lyraes, and if they were actually foreground RR Lyraes they would suggest the presence of a substantial and unexpected old stellar population in front of Carina at a mean distance of 55 kpc. However, Da Costa (1988) pointed out that the apparent magnitudes, amplitudes, and periods of at least four of these variables are consistent with their being anomalous Cepheids in Carina rather than foreground RR Lyraes. It is not yet known whether such variables occur in the Sextans system.

Analyses of the pulsational properties of the anomalous Cepheids (Norris and Zinn 1975; Demarque and Hirshfeld 1975; Hirshfeld 1980; Wallerstein and Cox 1984) indicated that they have masses of 1–2 $M_\odot$, about two or three times the mass of a typical RR Lyrae star. Theoretical models also indicated that variables in this mass range must be quite metal-poor, with [Fe/H] less than or equal to −1.5, in order to enter the instability strip. The higher masses of these stars compared to the RR Lyraes have brought forth two alternative explanations for their presence in dwarf spheroidal systems: either they are young stars aged 1–3 Gyr, implying that star formation has occurred at least that recently, or they are the result of mass transfer in older binary systems (Renzini, Mengel, and Sweigart 1977). Because at least one anomalous Cepheid has been detected in a globular cluster, V19 in NGC 5466 (Zinn and King 1982), at least some anomalous Cepheids are probably old.

Similar variables with period under one day have also been discovered in the Small Magellanic Cloud (Darteyet and Dessy 1952; Graham 1975; Smith et al. 1992). In the SMC, however, many of the 'bright RR Lyrae stars' with periods under one day are very short-period classical Cepheids, often pulsating in the first overtone mode. There is no reason for thinking that such stars are anything other than relatively young, metal-poor, single stars (Smith and Stryker 1986; Smith et al. 1992). With one or two possible exceptions (Graham 1985; Smith 1985b; Connolly 1985) no variables similar to anomalous Cepheids have been discovered in the LMC. This difference between the Clouds may be a reflection of their different age–metallicity relations: over the past few Gyr the LMC has been more metal-rich than the SMC (Smith and Stryker 1986).

Saha et al. (1992a) detected several variables in NGC 205, a dwarf elliptical companion to the Andromeda Galaxy, which may be anomalous Cepheids. A few possible anomalous Cepheids have been identified in the galactic field, but in the absence of clear luminosity information it is not easy to distinguish short-period galactic field anomalous Cepheids from RR Lyrae stars. Teays and Simon (1985) and Schmidt et al. (1990) have argued, however, that XZ Ceti may be a variable of this type.

Nemec, Wehlau, and Oliveira (1988) and Nemec (1989) have recently reanalyzed the period–luminosity relation for anomalous Cepheids. They found that these variables tend to fall along two distinct lines in the *P–L* diagram. They suggested that this separation represents a division of the anomalous Cepheids into fundamental mode and first-overtone mode pulsators. The period–luminosity relations for some well-studied anomalous Cepheids are shown in figure 6.6. It is not yet clear, however, whether all anomalous Cepheids fall neatly along the two sequences shown in figure 6.6. The Cepheids in Carina, for example, as well as some possibly similar variables in the SMC, show considerable scatter in the period–luminosity diagram.

6.4   The Andromeda Galaxy (M31) and its companions

At the beginning of this chapter, I noted that RR Lyrae stars in the Andromeda Galaxy perhaps played their most important role by initially not being seen. The story of the recognition that type I and type II Cepheids do not obey the same period–luminosity relation has been well-told elsewhere (Baade 1956; Fernie 1969), but a reprise emphasizing the role of RR Lyrae stars is worthwhile.

In the early fall of 1950 Walter Baade began to use the recently completed 5-m telescope at Palomar Mountain to take plates of three fields in the Andromeda Galaxy. His search was motivated by his doubts concerning the validity of the then accepted period–luminosity relation for Cepheids. That period–luminosity relation was a slightly

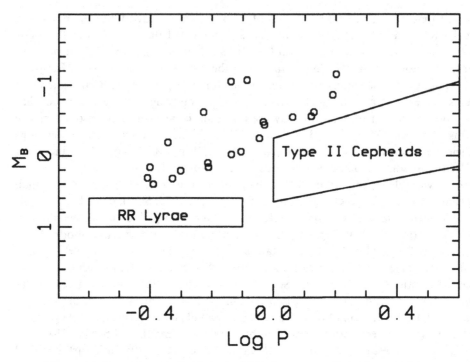

Figure 6.6. Anomalous Cepheids in the UMi, Draco, Leo II, Sculptor, NGC 5466 and Fornax dwarf spheroidal systems are plotted in the M_B vs. log period diagram. The anomalous Cepheids appear to fall along two period–luminosity lines. After Nemec et al. (1988).

revised version of the one Shapley had derived more than 30 years before. As discussed in §2.1, Shapley had established a calibration of the Cepheid period–luminosity relation based upon a statistical parallax solution for a small number of classical Cepheids in the Galaxy and Leavitt's period–luminosity relation for Cepheids in the Small Magellanic Cloud. Shapley then desired to tie the RR Lyrae stars into the Cepheid period-luminosity relation. At the time that Shapley was carrying out his investigations, *c.* 1916–1918, the data needed to determine a reliable statistical parallax from RR Lyrae stars alone were lacking. However, Shapley noted that some globular clusters, such as ω Centauri, contain both Cepheids and RR Lyrae stars. The globular cluster Cepheids could thus be used to link the RR Lyrae variables to the Cepheid period–luminosity relation. Implicit in that linkage was the assumption that Cepheids in the globular clusters obey the same period-luminosity relation as the Cepheids which Shapley had used to calibrate his period–luminosity relation. Shapley's study was completed long before Baade's discovery of stellar populations and his adoption of a universal period-luminosity relation for all Cepheids was a reasonable one. By this method, he arrived at the conclusion that the RR Lyrae stars had an absolute photographic magnitude near 0.0 and an absolute visual magnitude near −0.1.

When applied to Andromeda, however, this period–luminosity relation led to worrisome discrepancies. It made the upper limit in luminosity of globular clusters in

Andromeda about 1.5 magnitudes fainter than the same limit in the Galaxy. Baade (1956) described how he and Hubble debated among themselves the reason for this discrepancy during the 1940s, without arriving at a convincing explanation. When, later in that decade, Baade realized that stellar populations could be broadly divided into two basic types, the now well-known populations I and II, the idea of a universal period–luminosity relation seemed questionable to him. The Magellanic Cloud Cepheids studied by Leavitt and the classical Cepheids of the Galaxy belong to the young Population I. It was also upon observations of Population I classical Cepheids that the then accepted distance to the Andromeda Galaxy was based. Most RR Lyrae variables, on the other hand, and Cepheids found in globular clusters belong to Population II. Baade knew that Population I and Population II Cepheids differed in lightcurve shape, spectral characteristics, and period distribution (Baade 1951). Perhaps, Baade speculated, Shapley erred in linking the RR Lyrae stars with Population I Cepheids by means of globular cluster Cepheids of Population II.

Baade (1956) recounted: 'This was the situation when the 200-inch telescope was nearing completion and we were discussing the first observing programs for the new instrument. Naturally, I was very eager to settle these distubing questions which had arisen regarding the accepted period–luminosity relation.' The Andromeda Galaxy, he concluded, provided the most suitable northern object for a direct comparison of the two stellar populations. The 200-inch telecope was needed because RR Lyrae stars in Andromeda were too faint to be photographed with the Mt. Wilson 100-inch telescope, the world's largest until the Palomar instrument went into operation.

The plates Baade took led to a surprising result: 'Already the very first plates indicated that the accepted form of the period–luminosity relation did not represent the true situation'. If the accepted period–luminosity relation held, then Baade expected that RR Lyrae stars in Andromeda would appear at $m_{pg} = 22.4$, near the plate limit. Instead, only the brightest Population II stars were visible at that limit. Since in globular clusters such stars are about 1.5 magnitudes brighter photographically than the RR Lyrae stars, Baade concluded that the accepted period–luminosity relation made the RR Lyrae stars about 1.5 magnitudes too bright compared to the Population I Cepheids. Hence, the Andromeda RR Lyrae stars were really to be expected near an apparent photographic magnitude of 23.9, rather than 22.4, and so were beyond the limiting magnitude of the Palomar telescope.

Baade realized that the period–luminosity relations for Population I and Population II Cepheids must therefore differ, with the type I classical Cepheids averaging about 1.5 magnitudes brighter at a given period than their type II globular cluster counterparts. He set about determining the zero-points appropriate to each period–luminosity relation. He based the new zero-points on statistical parallax results for the absolute magnitudes of the RR Lyrae stars, most of which then still gave values near $M_{pg} = 0.0$. At the 1952 Rome meeting of the International Astronomical Union, he presented his conclusions, announcing that distances derived from type I Cepheids had to be multiplied by a factor 2, doubling the extragalactic distance scale. After the talk in which he presented these results, A. D. Thackeray rose to announce that he and A. J. Wesselink had discovered RR Lyrae stars in the SMC star cluster NGC 121. These variables were found at 19th magnitude, rather than 17th, as expected from the old period–luminosity relation, providing immediate confirmation of Baade's results.

The Andromeda RR Lyraes, after their dramatic nonappearance on Baade's

Palomar photographs, long escaped detection. With the addition of CCD detectors to large telescopes in locations of good seeing, that which eluded Baade has at last been realized. Pritchet and van den Bergh (1987) used the Canada–France–Hawaii telescope to take deep B CCD frames of a field on the SE minor axis of M31. This field lies about 40 arcminutes from the nucleus, which is equivalent to a projected distance of about 9 kpc. Within this field, they searched for RR Lyrae stars and were successful. They detected 32 suspected RR Lyrae stars and obtained preliminary periods for 28 of these, all ab-type variables. With their adopted value of $\langle M_B \rangle = +1.03 \pm 0.14$, their observed mean B magnitude of 25.68 ± 0.06, and $A_B = 0.31$, they determined the distance modulus of M31 to be 24.34 ± 0.15, corresponding to a distance of 740 ± 50 kpc.

The mean period for the RRab stars in Andromeda, $\langle P_{ab} \rangle = 0.548$, is similar to that of RRab stars in Oosterhoff I globular clusters. Pritchet and van den Bergh suspected, however, that their inventory of RR Lyraes in this field was seriously incomplete and that the true number of variables with amplitudes greater than 0.7 mag may be as high as 120 (note, however, that Saha et al. 1992a suggest that Pritchet and van den Bergh may have overestimated incompleteness). This would make the frequency of RR Lyraes in the M31 field similar to that of an RR Lyrae-rich globular cluster.

This relative abundance of RR Lyraes raised questions about the progenitors of the M31 RR Lyraes. Mould and Kristian's (1986) color–magnitude diagram for red giants in the M31 halo indicated that most are relatively metal-rich, with $\langle [Fe/H] \rangle = -0.6$. In the Galaxy, globular clusters with $[Fe/H] > -0.8$ generally have stubby red horizontal branches which do not extend far enough to the blue to enter the instability strip. As a consequence, the metal-rich globular clusters contain very few RR Lyrae stars. On the basis of the Mould and Kristian results, Pritchet and van den Bergh speculated that the M31 halo contained a relatively metal-rich, intermediate-to-blue horizontal branch population largely missing in the Galaxy. Subsequently, however, they derived a color–magnitude diagram for the red giants in their M31 field which indicated that the bulk of the stars might be slightly more metal-poor than found by Mould and Kristian (Pritchet and van den Bergh 1988). This led them to suggest that the globular cluster NGC 6171, with $[Fe/H] = -1.0$, might provide a good model of a population in the Galaxy comparable to that in the M31 halo. There is other evidence that some RR Lyrae stars in the Pritchet and van den Bergh field are only modestly metal deficient. Though much additional work is needed before definitive amplitudes and periods can be established for these variables, their preliminary locations in the period–amplitude diagram (figure 6.7) are consistent with a metallicity near -1 or above – if the relationship between amplitude, period, and metallicity which holds for most RR Lyraes in the Galaxy can be equally applied to those in M31. If the NGC 6171 analogy holds, then the RR Lyraes in the M31 field may be redder than assumed in Pritchet and van den Bergh's discovery paper, slightly reducing their derived distance modulus to $(m-M)_0 = 24.23 \pm 0.15$.

The techniques which finally succeeded in revealing RR Lyrae stars in M31, CCD photometry with a large telescope under conditions of good seeing, have also been successfully applied to three companion galaxies of M31, the dwarf elliptical galaxies NGC 147, NGC 185, and NGC 205. Two of these systems, NGC 147 and NGC 185, were previously considered to consist mainly, if not completely, of old stars. They contain no bright blue main sequence stars, nor areas of dense interstellar material. NGC 205, on the other hand, seems to contain a mixture of populations. Its central

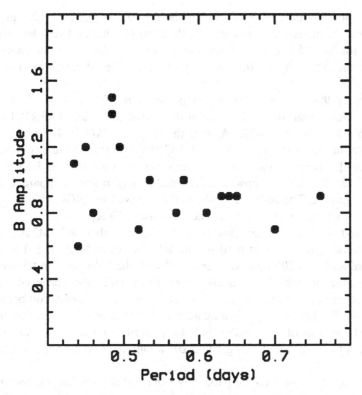

Figure 6.7 The RR Lyrae stars in M31 discovered by Pritchet and van den Bergh (1987) are plotted in the period–amplitude diagram. Variables which Pritchet and van den Bergh indicated as having uncertain amplitudes are not included.

region contains blue main sequence stars typical of Population I, and there are several patches of high interstellar absorption. However, NGC 205 also contains old red giants typical of Population II.

Saha and Hoessel (1987) used the 4-m telescope at Kitt Peak National Observatory to search for faint variable stars in NGC 147. Despite serious problems because of crowding in their field, which was centered 2.8 arcmin from the galaxy center, they detected 34 variable stars. Lightcurves could be determined for only 13 of these variables; nine of the 13 proved to be probable RR Lyrae stars. One anomalous Cepheid may have been detected.

Saha, Hoessel, and Mossman (1990), using deep observations obtained with the Palomar 5-m telescope in the g-passband (4930 Å) of Thuan and Gunn (1976), expanded this work to include a less crowded field about 6 arcmin northwest of the center of NGC 147. In this field they discovered 36 variable stars, 32 of which they believed to be probable RR Lyrae variables. Their results, though subject to some confusion among aliases of the RR Lyrae periods, indicated that NGC 147 RR Lyrae stars have a wide range in periods. A broad range in periods can sometimes be suggestive of a range in metal abundance, as in the globular cluster ω Centauri (see chapter 3). This led Saha et al. to suggest that the NGC 147 RR Lyraes may range significantly in metallicity. Adopting $\langle M_g \rangle = 0.73 \pm 0.25$ and $A_g = 0.60$, they estimated the distance modulus of NGC 147 to be $(m–M)_0 = 23.92 \pm 0.25$, giving NGC 147 a distance modulus somewhat smaller than the Pritchet and van den Bergh distance modulus for M31. Saha et al. pointed out, however, that the significance of this difference depends critically on the accuracy of their adopted extinction toward NGC 147.

Saha and Hoessel (1990) discovered 176 variable stars in NGC 185, of which 151 were believed to be RR Lyrae stars. They found the period distribution of RR Lyrae stars in NGC 185 also to be very wide. As was the case with NGC 147, this period width may indicate that the RR Lyrae stars in NGC 185, like those in ω Cen, have a wide range in [Fe/H]. From the data in hand, however, it could not be determined whether the NGC 185 RR Lyraes showed an unusually large range in apparent, and hence absolute, magnitude. The distance modulus they derived for NGC 185 was $(m–M)_0 = 23.79 \pm 0.25$, corresponding to a distance of about 570 Kpc.

Saha, Hoessel, and Krist (1992a) searched for RR Lyrae variables in NGC 205 with observational material similar to that used to find RR Lyraes in NGC 185. Deep g-passband images of an NGC 205 field were obtained with the Palomar 5-m telescope. With this material, they identified 54 variable stars, 30 of which they believed to be definitely RR Lyrae variables. In the case of NGC 205, there is a complication because some of the observed RR Lyraes might belong actually to the halo of M31, rather than to NGC 205 itself. Saha et al. concluded that the majority of the RR Lyraes they discovered belonged to NGC 205, placing NGC 205 about 100 kpc more distant than M31.

One noteworthy aspect of the series of papers by Saha and collaborators is the great care taken to analyze the completeness of their discoveries of variable stars. They modeled the discovery probability as a function of period, which makes it easier to see how the distribution of observations might affect the observed period distributions.

In comparing the frequency of RR Lyrae variables in M31 and its companions, Saha and Hoessel (1990) defined the parameter

$$\mu = N(GB)/N(RRL)$$

where $N(GB)$ is the number of giant branch stars that are 1.0 to 2.0 magnitudes brighter than the nominal mean magnitude of the RR Lyrae stars, and $N(RRL)$ is the number of RR Lyrae stars. Values of μ for systems analyzed by Saha and his collaborators are shown in Table 6.4. Saha et al.'s (1992b) values for the RR Lyrae-rich globular cluster M3 and the RR Lyrae-poor globular cluster 47 Tucanae are also indicated.

6.5   IC 1613

Saha et al. (1992b) again used deep CCD photometry with the Palomar 5-m telescope to discover 15 RR Lyrae stars in the Local Group dwarf irregular galaxy IC 1613. As with the studies of the companions to M31, the observations were carried out in the g-band of the Thuan and Gunn (1976) system. The discovery of these RR Lyrae stars indicated that IC 1613 has a very old stellar population, as well as the younger population represented by its classical Cepheids. Adopting $M_g = 0.73$, and an extinction of $A_g = 0.07$, Saha et al. derived a distance modulus of 24.10 ± 0.27 for IC 1613. Because IC 1613 contains both classical Cepheids and RR Lyrae stars, it is another test object for comparing the Population I and Population II distance scales.

Table 6.4. *Frequency of red giants to RR Lyraes*

System	Type	μ	References
M3	Glob. Cluster	6	Saha et al. 1992b
47 Tuc	Glob. Cluster	900	Saha et al. 1992b
Carina	Dwarf Sph.	15	Saha et al. 1990
NGC 147	Dwarf. Ellip.	53	Saha et al. 1990
NGC 185	Dwarf Ellip.	16	Saha and Hoessel 1990
M31 Halo	Sb	12–23	Pritchet and van den Bergh 1988 Saha et al. 1992a,b
IC 1613	Dwarf Irr.	40	Saha et al. 1992b

As with the LMC, there is some evidence for a small discrepancy between distances derived by the two calibrators (see §2.9).

6.6 M33

The study of RR Lyrae stars in the halo of M33 is at an early stage compared to the situation in Andromeda. Pritchet and van den Bergh (Pritchet 1988) have identified seven likely RR Lyrae candidates, but have not yet been able to determine periods for them. From these stars they obtained a distance modulus of 24.45 ± 0.2, but caution that this must be regarded as very preliminary.

6.7 Future prospects

Undoubtedly, the study of RR Lyrae stars in systems beyond the Milky Way is yet in its infancy, and much progress is to be expected in the next decade. With CCD detectors on 4m-class telescopes and with conditions of excellent seeing, RR Lyrae stars can now be detected to a distance of 1 Mpc. The newly repaired Hubble Space Telescope will produce images of faint, crowded star fields with a much higher resolution than ordinary ground-based telescopes. Pritchet (1988) estimated that, with the Hubble Space Telescope working to expectation, photometric observations of RR Lyrae stars will be feasible to a distance of about 3 Mpc.

We can thus expect the RR Lyrae stars to become increasingly important as primary distance indicators to nearby galaxies. We can also expect more intensive studies of RR Lyrae variables in nearer extragalactic systems. The ongoing effort to detect gravitational microlensing of star images in the direction of the Large Magellanic Cloud (Alcock et al. 1993) should, as a byproduct, produce a wealth of information about its variable stars, including its RR Lyraes. The main purpose of this investigation is to detect changes in the apparent brightness of LMC stars caused by gravitational lensing by massive compact halo objects (MACHOs) in the galactic halo. To do this requires the monitoring of the brightnesses of many LMC stars over relatively long time intervals, which should provide the most detailed information to date on many types of LMC variables. Similar investigations of galactic bulge stars are also expected to provide new information about RR Lyrae variables in the direction of the galactic center (Mateo 1993).

Glossary of symbols used in the text

B	Blue magnitude in the broadband photometric system of Johnson and Morgan; centered on 4400 Å
b	Galactic latitude
BC	Bolometric correction
[Fe/H]	Logarithm of the iron to hydrogen ratio of a star minus the logarithm of the iron to hydrogen ratio of the Sun
g	Surface gravity
I	Broadband near-infrared magnitude in the systems of Johnson (9000 Å) or Cousins (8100 Å)
l	Galactic longitude
$L_\odot$	A unit of one solar luminosity, 3.90×10^{33} erg/sec
m	Apparent magnitude
M	Absolute magnitude
m_{bol}	Bolometric magnitude
m_H	Infrared magnitude in a bandpass centered on 1.65 μm
m_K	Infrared magnitude in a bandpass centered on 2.2 μm
m_{pg}	Blue photographic magnitude in the International System
M_v	Absolute magnitude in the Johnson V photometric system
$M_\odot$	A unit of one solar mass, 1.99×10^{33} gm
P	Pulsation period
$\langle P_{ab} \rangle$	Mean period of RRab pulsators in a system
$\langle P_c \rangle$	Mean period of RRc pulsators in a system
Q	Pulsation constant from the pulsation equation (Ritter's relation)
R	Broadband red magnitude in the photometric systems of Johnson (7200 Å) or Cousins (6700 Å)
$R_\odot$	A unit of one solar radius, 6.96×10^8 m
RRab	RR Lyrae star of Bailey type a or b
RRc	RR Lyrae star of Bailey type c
RRd	Double mode RR Lyrae star
T_e	Effective temperature
U	Broadband near-ultraviolet magnitude in the photometric system of Johnson and Morgan; centered on 3650 Å
V	Broadband visual magnitude in the photometric system of Johnson and Morgan; centered on 5500 Å
X	Fraction of the mass of a star which consists of the element hydrogen
Y	Fraction of the mass of a star which consists of the element helium
Z	Fraction of the mass of a star which consists of elements heavier than helium
ZAHB	Zero-age horizontal branch
ΔS	Preston's (1959) spectroscopic metal abundance index for RR Lyrae stars
ρ	Density

Short list of journal abbreviations

A&A	*Astronomy and Astrophysics*
A&AS	*Astronomy and Astrophysics Supplements*
Acta Astron.	*Acta Astronomica*
AJ	*Astronomical Journal*
ApJ	*Astrophysical Journal*
ApJL	*Astrophysical Journal Letters*
ApJS	*Astrophysical Journal Supplement*
ARAA	*Annual Reviews of Astronomy and Astrophysics*
Astron. Nachr.	*Astronomische Nachrichten*
BAN	*Bulletin of the Astronomical Institutes of the Netherlands*
IBVS	*Information Bulletin on Variable Stars*
JAAVSO	*Journal of the American Association of Variable Star Observers*
Mem. S. A. It.	*Memorie della Societa Astronomica Italiana*
MNRAS	*Monthly Notices of the Royal Astronomical Society*
PASP	*Publications of the Astronomical Society of the Pacific*
PZ	*Peremeniye Zvezdy*

Bibliography

Gilmore, G., King, I. R., and van der Kruit, P. C. 1990, *The Milky Way as a Galaxy*, (Mill Valley: Univ. Science Books).

Hazen, M. L. 1986, *JAAVSO* **15**, 201.

Hoffmesiter, C., Richter, G., and Wenzel, W. 1985, *Variable Stars*, transl. by S. Dunlap, (Berlin: Springer-Verlag).

Kadla, Z. I., and Gerashchenko, A. N. 1984, *Izvestia Glavnoi Astronomicheskoi Observatorii b Pulkova*, No. 202, 83.

Kukarkin, B. V. 1975, in *Variable Stars and Stellar Evolution*, ed. V. E. Sherwood and L. Plaut, (Dordrecht: D. Reidel), p. 511.

Ledoux, P. and Walraven, T. 1958, *Handbuch der Physik*, ed. S. Flugge, (Berlin: Springer-Verlag) vol. LI, p. 353.

Mihalas, D. and Binney, J. 1981, *Galactic Astronomy*, second edition, (San Francisco: Freeman).

Preston, G. W. 1964, *ARAA* **2**, 23.

Rosino, L. 1973, in *Variable Stars in Globular Clusters and Related Systems*, ed. J. D. Fernie (Dordrecht: D. Reidel), p. 51.

Rosino, L. 1978, *Vistas in Astronomy*, **22**, 39.

Rowan-Robinson, M. 1985, *The Cosmological Distance Ladder*, (New York: Freeman).

Straizys, V. 1982, *Metal-Deficient Stars*, (Vilnius).

Tsesevich, V. P. 1966, *The RR Lyrae Stars*, (Kiev: Naukova Dumka).

Tsesevich, V. P. 1969, *The RR Lyrae Stars*, NASA, US Dept. of Commerce (English translation of Tsesevich 1966).

Tsesevich, V. P. 1975, in *Pulsating Stars*, ed. B. V. Kukarkin (New York: John Wiley), p. 144.

Woolley, R. 1965, *Quarterly Journal of the Royal Astron. Soc.* **6**, 2.

References

Abt, H. 1959, *ApJ* **130**, 824.

Abt, H. 1983, *ARAA* **21**, 343.

Alcock, C. et al. 1993, *Nature* **365**, 621.

Alexander, J. 1960, *MNRAS* **121**, 97.

Andrews, P. J. 1980, in *Star Clusters*, ed. J.E.Hesser, (Dordrecht: D. Reidel), p. 425.

Armandroff, T.E. and Zinn, R. 1988, *AJ* **96**, 92.

Arp, H.C. 1955, *AJ* **60**, 317.

Baade, W. 1926, *Astron. Nachr.* **228**, 359.

Baade, W. 1944, *ApJ* **100**, 137.

Baade, W. 1951, in *The Structure of the Galaxy*, U. Michigan Observ. Publ. **10**, 7.

Baade, W. 1956, *PASP* **68**, 5.

Baade, W. 1963, *Evolution of Stars and Galaxies*, ed. C. Payne-Gaposchkin (Cambridge: Harvard University Press).

Baade, W. and Hubble, E. P. 1939, *PASP* **51**, 40.

Baade, W. and Swope, H.H. 1961, *AJ* **66**, 300.

Babcock, H.W. 1955, *PASP* **67**, 70.

Babcock, H.W. 1958, *ApJS* **3**, 141.

Bailey, S.I. 1902, *Harv. Coll. Observ. Annals,* **38**, 1.

Bailey, S.I. 1913, *Harv. Coll. Observ. Annals,* **78**, No. 1.

Bailey, S.I. 1917, *Harv. Coll. Observ. Annals,* **78**, No. 2.

Baker, N.H. and Kippenhahn, R. 1965, *ApJ* **142**, 868.

Balazs-Detre, J. 1959, Kleine Veroff. Remeis-Sternw., No. 27, p. 26.

Balazs-Detre, J. and Detre, L. 1965, in *The Position of Variable Stars in the Hertzsprung–Russell Diagram*, Veroff. der Remeis-Sternwarte Bamberg, IV, Nr. 40, p. 184

Barlai, K. 1984, in *Observational Tests of Stellar Evolution Theory*, ed. A. Maeder and A. Renzini (Dordrecht: D. Reidel), p. 457.

Barnes, T.G., III and Hawley, S.L. 1986, *ApJL* **307**, L9.

Beers, T. and Sommer-Larsen, J. 1993, preprint.

Belserene, E.P. 1964, *AJ* **69**, 475.

Belserene, E.P. 1973, in *Variable Stars in Globular Clusters and Related Systems*, ed. J.D. Fernie, (Dordrecht: D. Reidel), p. 105.

Bingham, E.A., Cacciari, C., Dickens, R.F., and Fusi Pecci, F. 1984, *MNRAS* **209**, 765.

Blackwell, D.E. and Shallis, M.J. 1977, *MNRAS* **180**, 177.

Blanco, B. M.1984, *AJ* **89**, 1836.

Blanco, B.M. 1992, *AJ* **103**, 1872.

Blanco, V. 1992, *AJ* **104**, 734.

Blanco, V. and Blanco, B. 1985, *Mem. S.A.It.* **56**, 15.

Blaauw, A. 1955, in *Coordination of Galactic Rsearch*, IAU Symp. 1, (Cambridge: Cambridge University Press), p. 9.

Blazhko, S. 1907, *Astron. Nachr.* **175**, 325.

Bok, P.F. and Boyd, C.D. 1933, *Harv. Obs. Bull.* No. 893, 1.

Bonnell, J. T. and Bell, R.A. 1985, *PASP* **97**, 236.

Bonnell, J., Wu, C.-C., Bell, R.A., and Hutchinson, J.L. 1982, *PASP* **94**, 910.

Bookmeyer, B.B., Fitch, W.S., Lee, T.A., Wisniewski, W.Z., and Johnson, H.L., 1977, *Rev. Mex. Astron. Astrofis.* **2**, 235.

Borkowski, K.J. 1980, *Acta Astron.* **30**, 393.

Breger, M. 1990, in *Confrontation Between Stellar Pulsation and Evolution*, ed. C. Cacciari and G. Clementini, (San Francisco: ASP), p. 263.

Buonanno, R., Corsi, C.E., Fusi Pecci, F., Hardy, E., and Zinn, R. 1985, *A&A* **152**, 65.

Buonanno, R., Corsi, C.E., and Fusi Pecci, F. 1989, *A&A* **216**, 80.

Buonanno, R., Cacciari, C., Corsi, C.E., and Fusi Pecci, F. 1990, *A&A* **230**, 315.

Burki, G. and Meylan, G. 1986, *A&A* **156**, 131.

Burstein, D. and Heiles, C. 1982, *AJ* **87**, 1165.

Butler, D. 1975, *ApJ* **200**, 68.

Butler, D. and Deming, D. 1979, *AJ* **84**, 86.

Butler, D., Carbon, D., and Kraft, R.P. 1976, *ApJ* **210**, 120.

Butler, D., Dickens, R.J., and Epps, E. 1978, *ApJ* **225**, 148.

Butler, D., Kinman, T.D., and Kraft, R.P. 1979, *AJ* **84**, 993.

Butler, D., Kemper, E., Kraft, R.P., and Suntzeff, N.B. 1982a, *AJ* **87**, 353.

Butler, D., Manduca, A., Deming, D., and Bell, R.A. 1982b, *AJ* **87**, 640.

Butler, D., Demarque, P., and Smith, H.A. 1982c, *ApJ* **257**, 592.

Butler, D., Laird, J.B., Eriksson, K., and Manduca, A. 1986, *AJ* **91**, 570.

Cacciari, C. 1984, *AJ* **89**, 231.

Cacciari, C. and Renzini, A. 1976, *A&AS* **25**, 303.

Cacciari, C., Clementini, G., and Buser, R. 1989a, *A&A* **209**, 141

Cacciari, C., Clementini, G., Prevot, L., and Buser, R. 1989b, *A&A* **209**, 154.

Cacciari, C., Clementini, G., and Fernley, J.A. 1992, *ApJ* **396**, 219.

Caputo, F. 1985, *Rep. Prog. Phys.* **48**, 1235.

Caputo, F. 1988, *A&A* **189**, 70.

Caputo, F. and De Santis, R. 1992, *AJ* **104**, 253.

Carney, B.W. 1980, *ApJS* **42**, 481.

Carney, B.W. and Jones, R. 1983, *PASP* **95**, 246.

Carney, B.W. and Latham, D.W. 1984, *ApJ* **278**, 241.

Carney, B.W. and Seitzer, P. 1986, *AJ* **92**, 23.

Carney, B.W., Storm, J., and Jones, R.V. 1992, *ApJ* **386**, 663.

Carney, B.W., Storm, J., Trammell, S.R., and Jones, R.V. 1992, *PASP* **104**, 44.

Catelan, M. 1992, *A&A* **261**, 457.

Castellani, V. 1983 *Mem. S. A. It.* **54**, 141.

Castellani, V. 1985, *Fund. of Cosmic Phys.* **9**, 317.

Castellani, V. and Quarta, M.L. 1987, *A&AS* **71**, 1.

Christy, R.F. 1966, *ApJ* **144**, 108.

Clement, C.M. 1990, *AJ* **99**, 240.

Clement, C.M., Nemec, J.M., Robert, N., Wells, T., Dickens, R.J., and Bingham, E.A. 1986, *AJ* **92**, 825.

Clement, C.M. and Nemec, J.M. 1990, *JRASC* **84**, 424.

Clement, C.M. and Walker, I.R. 1991, *AJ* **101**, 1352.

Clement, C.M., Kinman, T.D., and Suntzeff, N.B. 1991, *ApJ* **372**, 273.

Clement, C.M., Jankulak, M., and Simon, N.R. 1992, *ApJ* **395**, 192.

Clement, C.M., Ferance, S., and Simon, N.R. 1993, *ApJ* **412**, 183.

Clementini, G., Tosi, M., and Merighi, R. 1991, *AJ* **101**, 2168.

Clube S.V.M. and Dawe, J.A. 1980, *MNRAS* **190**, 575.

Cohen, J.G. 1992, *ApJ* **400**, 528.

Cohen, J.G. and Gordon, G.A. 1987, *ApJ* **318**, 215.

Connolly, L. P. 1985, *ApJ* **299**, 729.

Cousens, A. 1983, *MNRAS* **203**, 1171.

Coutts, C.M. 1971, in *New Directions and New Frontiers in Variable Star Research*, Veroff. der Remeis-Sternwarte Bamberg, IX, Nr. 100, 238.

Coutts, C.M. and Sawyer Hogg, H. 1969, *Publ. David Dunlap Obs.* **3**, 1.

Coutts, C.M. and Sawyer Hogg, H. 1971, *Publ. David Dunlap Obs.* **3**, 61.

Cox, A.N. 1980, *Space Sci. Rev.* **27**, 475.

Cox, A.N. 1987, in *Second Conference on Faint Blue Stars*, ed. A.G.D. Philip, D.S. Hayes, and J.W. Leibert (Schenectady: L. Davis Press), p. 161.

Cox, A.N. 1993a, in *New Perspectives on Stellar Pulsation and Pulsating Variable Stars*, ed. J.M. Nemec and J.M. Matthews (Cambridge: Cambridge Univ. Press), p. 409.

Cox, A.N. 1993b, in *New Perspectives on Stellar Pulsation and Pulsating Variable Stars*, ed. J.M. Nemec and J.M. Matthews (Cambridge: Cambridge Univ. Press), p. 241.

Cox, A.N., King, D.S., and Hodson, S.W. 1978, *ApJ* **224**, 607.

Cox, A.N., King, D.S., and Hodson, S.W. 1980, *ApJ* **236**, 219.

Cox, A.N., Hodson, S.W., and Clancy S.P. 1983, *ApJ* **266**, 94.

Cox, J.P. 1974, *Rep. Prog. Phys.* **37**, 563.

Cox, J.P. 1980, *Theory of Stellar Pulsation*, (Princeton: Princeton Univ. Press).

Cox, J.P. 1985, in *Cepheids: Theory and Observations*, ed. B.F. Madore, (Cambridge: Cambridge Univ. Press), p. 126.

Cox, J.P. and Whitney, C.A. 1958, *ApJ* **127**, 561.

Cudworth, K. and Peterson, R. C. 1988, in *Globular Cluster Systems in Galaxies*, ed. J.E. Grindlay and A.G.D. Philip (Dordrecht: Kluwer), p. 523.

Da Costa, G.S. 1984, *ApJ* **285**, 483.

Da Costa, G.S. 1988, in *Globular Cluster Systems in Galaxies*, ed. J.E. Grindlay and A.G.D. Philip (Dordrecht: Kluwer), p. 217.

Da Costa, G.S. and Armandroff, T.E. 1990, *AJ* **100**, 162.

Dartayet, M. and Dessy, J.L. 1952, *ApJ* **115**, 279.

Dearborn, D., Raffelt, G., Salat, P., Silk, J., and Bouquet, A. 1990, *ApJ* **354**, 568.

Demarque, P. and Hirshfeld, A.W. 1975, *ApJ* **202**, 346.

Demers, S. and Kunkel, W.E. 1976, *ApJ* **208**, 932.

Demers, S. and Wehlau, A. 1977, *AJ* **82**, 620.

Detre, L. 1969, in *Non-Periodic Phenomena in Variable Stars*, ed. L. Detre (Dordrecht: D. Reidel), p. 3.

Detre, L. and Szeidl, B.1973, in *Variable Stars in Globular Clusters and Related Systems*, ed. J.D. Fernie (Dordrecht: D. Reidel), p. 31.

Deupree, R.G. 1977a, *ApJ* **211**, 509.

Deupree, R.G. 1977b, *ApJ* **214**, 502.

Dickens, R.J. 1989, in *The Use of Pulsating Stars in Fundamental Problems in Astronomy*, ed. E.G. Schmidt, (Cambridge: Cambridge University Press), p. 141.

Diethelm, R. 1983, *A&A* **124**, 108.

Diethelm, R. 1986, *A&AS* **64**, 261.

Dorman, B. 1992, *ApJS* **81**, 221.

Dorman, B., Lee, Y.-W., and VandenBerg, D.A. 1991, *ApJ* **366**, 115.

Eddington, A.S. 1918, *MNRAS* **79**, 2.

Eddington, A.S.1926, *The Internal Constitution of the Stars*, (Cambridge: Cambridge Univ.Press).

Eggen, O.J., Lynden-Bell, D., and Sandage, A. R. 1962, *ApJ* **136**, 748.

Epstein, I. 1950, *ApJ* **112**, 6.

Epstein, I. 1969, *AJ* **74**, 1131.

Epstein, I. and Epstein, A.E.A. 1973, *AJ* **78**, 83.

Feast, M.W. and Walker, A.R. 1987, *ARAA* **25**, 345.

Fernie, J.D. 1969, *PASP* **81**, 707.

Fernley, J.A., Longmore, A.J., Jameson, R.F., Watson, F.G., and Wesselink, T. 1987, *MNRAS* **226**, 927.

Fernley, J.A., Lynas-Gray, A.E., Skillen, I., Jameson, R.F., Marang, F., Kilkenny, D., and Longmore, A.J. 1989, *MNRAS* **236**, 447.

Fernley, J.A., Skillen, I., Jameson, R.F., and Longmore, A.J. 1990a, *MNRAS* **247**, 287.

Fernley, J.A., Skillen, I., Jameson, R.F., and Longmore, A.J. 1990b, *MNRAS* **242**, 685.

Firmaniuk, B.N., Kreiner, J.M., and Zakrzewski, B. 1988, in *Rocznik Astron. Obsew. Krakowskiego*, ed. K. Rudnicki, Nr. 60.

Fitch , W.S. 1967, *ApJ* **148**, 481.

Fitch, W.S. and Szeidl, B. 1976, *ApJ* **203**, 616.

Fitch, W.S., Wisniewski, W.Z., and Johnson, H.L. 1966, *Comm. Lunar & Planetary Lab.*, **5**, No. 71.

Fletcher, A. 1934, *MNRAS* **95**, 56.

Fokin, A.B. 1992, *MNRAS* **256**, 26.

Freeman, K.C. and Rodgers, A.W. 1975, *ApJL* **201**, L71.

Fusi Pecci, F., Ferraro, F.R., Crocker, D.A., Rood, R.T., and Buonanno, R. 1990, *A&A* **238**, 95.

Gascoigne, S.C.B. 1966 *MNRAS* **134**, 59.

Gehmeyr, M. 1993, *ApJ* **412**, 341.

Geyer 1973, in *Variable Stars in Globular Clusters and Related Systems*, ed. J.D. Fernie (Dordrecht: D. Reidel), p. 88.

Gillet, D. and Crowe, R.A. 1988, *A&A* **199**, 242.

Gillet, D., Burki, G., and Crowe, R.A. 1989, *A&A* **225**, 445.

Gilmore, G., King, I.R., and van der Kruit, P.C. 1990, *The Milky Way as a Galaxy*, (Mill Valley: Univ. Science Books).

Gloria, K.A. 1990, *PASP* **102**, 338.

Goldsmith, C. G. 1993, in *New Perspectives on Stellar Pulsation and Pulsating Variable Stars*, ed. J.M. Nemec and J.M. Matthews (Cambridge: Cambridge Univ. Press), p. 358.

Goranskij, V.P. 1976, *PZ* **3**, No. 13.

Goranskij, V.P. 1981, *IBVS* No. 2007.

Goranskij, V.P. 1989, *Sov. Astron.* **33**, 45.

Goranskij, V.P. and Shugarov, S.Y. 1979, *PZ* **21**, 211.

Gordenko, A.F., Klabukova, A.V., and Fashchevskil, N.N. 1984, *Problemy Kosmicheskoj Fiziki* **19**, 93.

Graham, J.A. 1975, *PASP* **87**, 641.

Graham, J.A. 1977, *PASP* **89**, 425.

Graham, J.A. 1985, *PASP* **97**, 676.

Graham, J.A. and Nemec, J.M. 1984, in *Structure and Evolution of the Magellanic Clouds*, ed. S. van den Bergh and K.S. de Boer (Dordrecht: D. Reidel), p. 37.

Graham, J.A. and Ruiz, M.T. 1974, *AJ* **79**, 363.

Gratton, R.G. and Ortolani, S. 1987, *A&AS* **71**, 131.

Gratton, R.G., Tornambe, A., and Ortolani, S. 1986, *A&A* **169**, 111.

Grosse, E. 1932, *Astron. Nachr.* **246**, 376.

Gryzunova, T.I. 1972, *Perem. Zvezdy Suppl.* **1**, No. 4.

Gryzunova, T.I. 1979, *Astron. Tsirk.* No. 1075, 7.

Hachenberg, O. 1939, *Zs.f.Ap.* **18**, 49.

Hanson, R.B. 1979, *MNRAS* **189**, 875.

Hardie, R.H. 1955, *ApJ* **122**, 256.

Hartwick, F.D.A. 1987, in *The Galaxy*, ed. G. Gilmore and B. Carswell, (Dordrecht: Reidel), p. 281.

Hartwick, F.D.A., Barlow, D.J., and Hesser, J.E. 1981, *AJ* **86**, 1044.

Hawkins, M.R.S. 1984, *MNRAS* **206**, 433.

Hawley, S.L., Jeffreys, W.H., Barnes, T.G. III, and Lai, Wan 1986, *ApJ* **302**, 626.

Hazen, M.L. and Nemec, J.M. 1992, *AJ* **104**, 111.

Heck, A. 1988. *Bull. Centre Donnees Stell.* **34**, 133.

Heck, A. and Lakaye, J.M. 1978, *MNRAS* **184**, 17.

Hemenway, M.K. 1975, *AJ* **80**, 199.

Hertzsprung, E. 1909, *Astron. Nachr.* **179**, 376.

Hertzsprung, E. 1913, *Astron. Nachr.* **196**, 201.

Hill, S.J. 1972, *ApJ* **178**, 793.

Hirshfeld, A.W. 1980, *ApJ* **241**, 111.

Hodge, P. W. and Wright, F.W. 1978, *AJ* **83**, 228.

Hoffleit, D. 1993, private communication.

Hoffmesiter, C., Richter, G., and Wenzel, W. 1985, *Variable Stars*, transl. by S. Dunlap, (Berlin: Springer-Verlag), pp. 46–47.

Hutchinson, J.L., Hill, S.J., and Lillie, C.F. 1977, *ApJ* **211**, 207.

Iben, I. Jr. 1968, *Nature* **220**, 143.

Iben, I. Jr. 1971, *PASP* **83**, 679.

Iben, I. Jr. and Rood, R.T. 1970, *ApJ* **161**, 587.

Iglesias, C.A., Rogers, F.J., and Wilson, B.G. 1990, *ApJ* **360**, 221.

Innanen, K.A. and Papp, K.A. 1979, *AJ* **84**, 601.

Iwanowska, W. 1953, *Bull. Torun Obs.* No. 11.

Jameson, R.F. 1986, *Vistas in Astron.* **29**, 17.

Jerzykiewicz, M., Schult, R.H., and Wenzel, W. 1982, *Acta Astron.* **32**, 357.

Jones, D.H.P. 1971, *MNRAS* **154**, 79.

Jones, D.H.P. 1973, *ApJS* **25**, 487.

Jones, R.V. 1988, *ApJ* **326**, 305.

Jones, R.V., Carney, B.W., Latham, D.W., and Kurucz, R.L. 1987a, *ApJ* **312**, 254.

Jones, R.V., Carney, B.W., Latham, D.W., and Kurucz, R.L. 1987b, *ApJ* **314**, 605.

Jones, R.V., Carney, B.W., and Latham, D.W. 1988a, *ApJ* **326**, 312.

Jones, R.V., Carney, B.W., and Latham, D.W. 1988b, *ApJ* **332**, 206.

Jones, R.V., Carney, B.W., Storm, J., and Latham, D.W. 1992, *ApJ* **386**, 646.

Joy, A.H. 1938, *PASP* **50**, 302.

Jurcsik, J. and Barlai, K. 1990, in *Confrontation Between Stellar Pulsation and Evolution*, ed. C. Cacciari and G. Clementini, (San Francisco: ASP), p. 112.

Kanyo, S. 1975, *Mitt. Ungar. Akad. Wiss.* No. 69.

Kanyo, S. 1980, *IBVS* No. 1832.

Kapteyn, J.C. 1890, *Astron. Nachr.* **125**, 169,

Kapteyn, J.C. and van Rhijn, P.J. 1922, *BAN* **1**, 37.

Kemper, E. 1982, *AJ* **87**, 1395.

Kholopov, P.N. et al. 1985 *General Catalogue of Variable Stars*, Fourth Edition, (Moscow).

Kiess C.C.1912, *PASP* **24**, 186.

King, I.R. 1966, *AJ* **71**, 64.

King, D.S. and Cox, J.P. 1968, *PASP* **80**, 365.

King, C.R., Demarque, P., and Green, E.M. 1988, in *Calibration of Stellar Ages*, ed. A.G.D. Philip (Schenectady: L. Davis Press), p. 211.

Kinman, T.D. 1959a, *MNRAS* **119**, 538.

Kinman, T.D. 1959b, *MNRAS* **119**, 559.

Kinman, T.D. and Carretta, E. 1992, *PASP* **104**, 111.

Kinman, T. D., Wirtanen, C.A., and Janes, K.A. 1965, *ApJS* **11**, 223.

Kinman, T.D., Wirtanen, C.A., and Janes, K.A. 1966, *ApJS* **13**, 379.

Kinman, T.D., Stryker, L.L., and Hesser, J.E. 1976, *PASP* **88**, 393.

Kinman, T.D., Mahaffey, C.T., and Wirtanen, C.A. 1982, *AJ* **87**, 314.

Kinman, T.D., Wong-Swanson, B., Wenz, M., and Harlan, E.A. 1984, *AJ* **89**, 1200.

Kinman, T. D., Kraft, R.P., Friel, E. and Suntzeff, N.B. 1985, *AJ* **90**, 95.

Kinman, T.D., Stryker, L.L., Hesser, J.E., Graham, J.A., and Walker, A.R. 1991, *PASP* **103**, 1279.

Klepikova, L.A. 1958, *PZ* **12**, 164.

Kluyver, M. 1936, *BAN* **7**, 313.

Koopmann, R.A., Lee, Y.-W., Demarque, P., and Howard, J.M. 1994, *ApJ* **423**, 380.

Kovacs, G., Buchler, J.R., and Marom, A. 1991, *A&A* **252**, L27.

Kovacs, G., Buchler, J.R., Marom, A., Iglesias, C.A., and Rogers, F.J., 1992, *A&A* **259**, L46.

Kraft, R.P. 1972, in *The Evolution of Population II Stars*, ed. A.G.D. Philip, Dudley Observ. Rept. No. 4, p. 69.

Kraft, R.P., Sneden, C., Langer, G.E., and Shetrone, M.D. 1993, *AJ* **106**, 1490.

Kuhn, J.R. 1993, *ApJL* **409**, L13.

Kuhn, J.R. and Miller, R.H. 1989, *ApJL*, **341**, L41.

Kukarkin, B.V. 1949, in *Structure and Evolution of Stellar Systems*, (Moscow: Government Printing Office), p. 182.

Kukarkin, B.V. 1973, in *Variable Stars in Globular Clusters and Related Systems*, ed. J.D. Fernie, (Dordrecht: D. Reidel), p. 8.

Kukarkin, B.V. 1975, in *Variable Stars and Stellar Evolution*, ed. V.E. Sherwood and L. Plaut, (Dordrecht: D. Reidel), p. 511.

Kukarkin, B.V. and Kukarkina, N. 1973, *PZ Suppl.* **1**, No. 1.

Kukarkin, B.V. and Kukarkina, N. 1980, *PZ* **21**, 365.

Kurtz, D.W. 1988, in *Multimode Stellar Pulsations*, ed. G. Kovacs, L. Szabados, and B. Szeidl (Budapest: Konkoly Observ.), p. 95.

Kurucz, R.L. 1979, *ApJS* **40**, 1.

Layden, A.C. 1993, in *The Globular Cluster–Galaxy Connection*, ed. G.H. Smith and J.P. Brodie, (San Francisco: ASP), p. 326.

Leavitt, H. 1908, *Harv. Coll. Observ. Annals* **60**, 87.

Lebre, A. 1993, in *New Perspectives on Stellar Pulsation and Pulsating Variable Stars*, ed. J.M. Nemec and J.M. Matthews (Cambridge: Cambridge Univ. Press), p. 376.

Ledoux, P. 1951, *ApJ* **114**, 373.

Ledoux, P. 1963 in *Stellar Evolution*, ed. L. Gratton (New York: Academic Press), p. 394.

Lee, M. G., Freedman, W., Mateo, M., Thompson, I., Roth, M., and Ruiz, M.-T. 1993, *AJ* **106**, 1420.

Lee, Y.W. 1990, *ApJ* **363**, 159.

Lee, Y.-W. 1991a, *ApJL* **373**, L43.

Lee, Y.-W. 1991b, *ApJ* **367**, 524.

Lee, Y.-W. 1992a, in *Variable Stars and Galaxies*, ed. B. Warner, (San Francisco: ASP), p. 103.

Lee, Y.-W. 1992b, *AJ* **104**, 1780.

Lee, Y.-W. 1992c, *PASP* **104**, 798.

Lee, Y.-W. and Demarque, P. 1990, *ApJS* **73**, 709.

Lee, Y.-W., Demarque, P., and Zinn, R. 1990, *ApJ* **350**, 155.

Liu, T. and Janes, K.A. 1989, *ApJS* **69**, 593.

Liu, T. and Janes, K.A. 1990a, *ApJ* **354**, 273.

Liu, T. and Janes, K.A. 1990b, *ApJ* **360**, 561.

Lloyd, C. 1992, in *Variable Star Research: An International Perspective*, ed. J. R. Percy, J.A. Mattei, and C. Sterken (Cambridge: Cambridge Univ. Press), p. 242.

Longmore, A.J. 1993, in *New Perspectives on Stellar Pulsation and Pulsating Variable Stars*, ed. J.M. Nemec and J.M. Matthews (Cambridge: Cambridge Univ. Press), p. 21.

Longmore, A.J., Fernley, J.A., Jameson, R.F., Sherrington, M.R., and Frank, J. 1985, *MNRAS* **216**, 873.

Longmore, A.J., Dixon, R., Skillen, I., Jameson, R.F., and Fernley, J.A. 1989, in *The Use of Pulsating Stars in Fundamental Problems in Astronomy*, ed. E.G. Schmidt, (Cambridge: Cambridge Univ. Press), p. 274.

Longmore, A.J., Dixon, R., Skillen, I., Jameson, R.F., and Fernley, J.A. 1990, *MNRAS* **247**, 684.

Lub, J. 1977, *A&AS* **29**, 345.

Lub, J. 1979, *AJ* **84**, 383.

Lub, J. 1987, in *Stellar Pulsation*, ed. A.N. Cox, W.M. Sparks, and S.G. Starrfield, (Berlin: Springer), p. 218.

Madore, B.F. and Freedman, W.L. 1991, *PASP* **103**, 933.

Manduca, A. 1981, *ApJ* **245**, 248.

Manduca, A. and Bell, R.A. 1981, *ApJ* **250**, 306.

Martin, W.C. 1938, *Ann. Sterrew. Leiden* **17**, part 2.

Mateo, M. 1993, *PASP* **105**, 1075.

Mateo, M., Nemec, J., Irwin, M., and McMahon, R. 1991, *AJ* **101**, 892.

Mathewson, D.S., Ford, V.L., and Visvanathan, N. 1986, *ApJ* **301**, 664.

McDonald, L.H. 1977, Ph.D. Dissertation, Univ. of California, Santa Cruz.

Mendes de Oliveira, C. and Nemec, J.M. 1988, *PASP* **100**, 217.

Menzies, J. 1974 , *MNRAS* **168**, 177.

Moffett, T.J. 1989, in *The Use of Pulsating Stars in Fundamental Problems in Astronomy*, ed E. G. Schmidt, (Cambridge: Cambridge Univ. Press), p. 191.

Morrison, H.L., Flynn, C., and Freeman, K.C. 1990, *AJ* **100**, 1191.

Moskalik, P. 1986, *Acta Astron.* **36**, 333.

Mould, J.R. and Aaronson, M. 1983, *ApJ* **273**, 530.

Mould, J.R. and Kristian, J. 1986, *ApJ* **305**, 591.

Munch, G. and Terrazas, L.R. 1946, *ApJ* **103**, 371.

Nemec, J.M. 1985a, *AJ* **90**, 240.

Nemec, J.M. 1985b, *AJ* **90**, 204.

Nemec, J.M. 1989, in *The Use of Pulsating Stars in Fundamental Problems in Astronomy*, ed. E.G. Schmidt, (Cambridge: Cambridge Univ. Press), p. 215.

Nemec, J.M. and Clement, C.M. 1989, *AJ* **98**, 860.

Nemec, J. M. and Mateo, M. 1990, in *Confrontation Between Stellar Pulsation and Evolution*, ed. C. Cacciari and G. Clementini, (San Francisco: ASP), p. 64.

Nemec, J.M., Hazen-Liller, M.L., and Hesser, J.E. 1985, *ApJS* **57**, 329.

Nemec, J.M., Hesser, J.E., and Ugarte, P. 1985, *ApJS* **57**, 387.

Nemec, J.M., Linnell Nemec, A.F. and Norris, J. 1986, *AJ* **92**, 358.

Nemec, J.M., Wehlau, A. and Mendes de Oliveira, C. 1988, *AJ* **96**, 528.

Nikolov, N., Buchantsova, N., and Frolov, M. 1984, *Catalog of RR Lyrae Photometric Properties: The Mean Light and Color Curves of 210 Field RR Lyrae Stars* (Sofia: Academy of Sciences).

Norris, J. 1986, *ApJS* **61**, 667.

Norris, J. and Zinn, R. 1975, *ApJ* **209**, 335.

Norris, M.V. 1973, in *Variable Stars in Globular Clusters and Related Systems*, ed. J.D. Fernie (Dordrecht: D.Reidel), 113.

Novikova, M.F. 1988, private communication.

O'Connell, D.J.K., 1958, *Stellar Populations*, (Specola Vaticana), Ric A, 5.

Oke, J.B. 1966, *ApJ* **145**, 468.

Oke, J.B., Giver, L.P., and Searle, L. 1962, *ApJ* **136**, 393.

Olszewski, E.W. 1988, in *Globular Cluster Systems in Galaxies*, ed. J.E. Grindlay and A.G.D. Philip (Dordrecht: Kluwer), p. 159.

Olszewski, E.W. and Aaronson, M. 1985, *AJ* **90**, 2221.

Olszewski, E.W., Schommer, R.A., and Aaronson, M. 1987, *AJ* **93**, 565.

Oort, J.H. and Plaut, L. 1975, *A&A* **41**, 71.

Oosterhoff, P. Th., 1939, *Observatory* **62**, 104.

Oosterhoff, P. Th. 1944, *BAN* **10**, 55.

Oosterhoff, P.Th. 1946, *BAN* **10**, 101.

Packer, D.E. 1890, *English Mechanic* **51**, 378.

Panagia, N., Gilmozzi, R., Machetto, F., Adorf, H.-M., and Kirshner, R.P. 1991, *ApJL* **380**, L23.

Pavlovskaya, E.D.1953, *PZ* **9**, 349.

Payne-Gaposchkin, C. and Gaposchkin, S. 1938, Variable Stars, (Cambridge: Harvard College Observ.).

Payne-Gaposchkin, C. and Gaposchkin, S. 1966, Smithsonian Contributions to Astrophysics. 9, 1.

Peniche, R., Gomez, T., Parrao, L., and Pena, J.H. 1989, A&A 209, 59.

Perek, L. 1951, Brno Contributions. 1, No. 8.

Petersen, J.O. 1973, *A&A* **27**, 89.

Petersen, J.O., 1990a, in *Confrontation Between Stellar Pulsation and Evolution*, ed. C. Cacciari and G. Clementini, (San Francisco: ASP), p. 402.

Petersen, J.O. 1990b, *A&A* **238**, 160.

Pickering, E. C. 1901, *Astron. Nachr.* **154**, 425.

Pickering, E.C. 1912, *Harv. Coll. Observ. Circ.* No. 173.

Plaut, L. 1966, *BAN* Suppl. 1, 105.

Plaut, L. 1968, *BAN* Suppl. 2, 293.

Plaut, L. 1971, *A&A* **8**, 341.

Plummer, H.C. 1913, *MNRAS* **73**, 665.

Prager, R. 1939, *Harv. Bull.* No. 911, 1.

Preston, G.W. 1959, *Ap.J.* **130**, 507.

Preston, G.W. 1961a, *ApJ* **133**, 29.

Preston, G.W. 1961b, *ApJ* **134**, 633.

Preston, G.W. 1964, *ARAA* **2**, 46.

Preston, G.W. 1967, in *The Magnetic and Related Stars*, (Baltimore: Mono Book Corp), p.3.

Preston, G.W. and Paczynski, B. 1964, *ApJ* **140**, 181.

Preston, G.W., Smak, J. and Paczynski, B. 1965, *ApJS* **12**, 99.

Preston, G.W., Shectman, S.A., and Beers, T.C. 1991, *ApJ* **375**, 121.

Pritchet, C.J. 1988, in *The Extragalactic Distance Scale*, ed. S. van den Bergh and C.J. Pritchet, (San Francisco: ASP), p. 59.

Pritchet, C.J. and van den Bergh, S. 1987, *ApJ* **316**, 517.

Pritchet, C.J. and van den Bergh, S. 1988, *ApJ* **331**, 135.

Rees, R.F., 1993, in *The Globular Cluster-Galaxy Connection*, ed. G.H. Smith and J.P. Brodie, (San Francisco: ASP), p. 104.

Reid, N. and Freedman, W. 1990, in *Confrontation Between Stellar Pulsation and Evolution*, ed. C. Cacciari and G. Clementini, (San Francisco: ASP), p. 120.

Renzini, A. 1977, in *Advanced Stages in Stellar Evolution*, ed. P. Bouvier and A. Maeder, (Sauverney: Geneva Observ.), p. 151.

Renzini, A. 1983, *Mem. SAIt.* **54**, 335.

Renzini, A., Mengel, J. and Sweigart, A.V. 1977, *A&A* **56**, 369.

Rich, R.M. 1988, *AJ* **95**, 828.

Roberts, M.S. and Sandage, A. 1955, *AJ* **60**, 185.

Rodgers, A.W. 1977, *ApJ* **212**, 117.

Romanov, Yu. S., Udovichenko, S.N., and Frolov, M.S. 1987, *Pis'ma Astron. Zh.* **13**, 69.

Rood, R.T. 1990, in *Confrontation Between Stellar Pulsation and Evolution*, ed. C. Cacciari and G. Clementini, (San Francisco: ASP), p. 11.

Russell, H.N. 1927, *ApJ* **66**, 122.

Saha, A. 1984, *ApJ* **283**, 580.

Saha, A. and Hoessel, J.G. 1987, *AJ* **94**, 1556.

Saha, A. and Hoessel, J.G. 1990, *AJ* **99**, 97.

Saha, A. and Oke, J.B. 1985, *ApJ* **285**, 688.

Saha, A. and White, R.E. 1990, *PASP* **102**, 148 (erratum: **102**, 495).

Saha, A., Hoessel, J.G. and Krist, J. 1992a, *AJ* **103**, 84.

Saha, A., Hoessel, J.G. and Mossman, A.E. 1990, *AJ* **100**, 108.

Saha, A., Monet, D.G. and Seitzer, P. 1986, *AJ* **92**, 302.

Saha, A., Freedman, W.L., Hoessel, J.G., and Mossman, A.E. 1992b, *AJ* **104**, 1072.

Sandage, A. 1958, in *Stellar Populations*, ed. D. O'Connell, (Specola Vaticana), Ric A, 5, 41.

Sandage, A. 1970, *ApJ* **162**, 841.

Sandage, A. 1981, *ApJ* **248**, 161.

Sandage, A. 1982a, *ApJ* **252**, 553.

Sandage, A. 1982b, *ApJ* **252**, 574.

Sandage, A. 1990a, *ApJ* **350**, 603.

Sandage, A. 1990b, *ApJ* **350**, 631.

Sandage, A. 1993a, *AJ* **106**, 687.

Sandage, A. 1993b, *AJ* **106**, 703.

Sandage, A. 1993c, *AJ* **106**, 719.

Sandage, A. and Cacciari, C. 1990, *ApJ* **350,** 645.

Sandage, A. and Wildey, R. 1967, *ApJ* **150**, 469.

Sandage, A., Katem, B., and Sandage, M. 1981, *ApJS* **46**, 41.

Sanford, R.1949, *ApJ* **109**, 208.

Sawyer, H. 1939, *David Dunlap Obs. Publ.* **1**, No. 4.

Sawyer, H. 1944, *Communications of David Dunlap Obs.* No. 11.

Sawyer, H. 1955, *David Dunlap Obs. Publ.* **2**, No. 4.

Sawyer Hogg, H. 1973, *David Dunlap Obs. Publ.* **3**, No. 6.

Schmidt, B.P., Kirshner, R.P., and Eastman, R.G. 1992, *ApJ* **395**, 366.

Schmidt, E.G., Loomis, C.G., Groebner, A.T., and Potter, C.T. 1990, *ApJ* **360**, 604.

Schwarzschild, M. 1940, *Harvard College Observ. Circ.* No. 437.

Schwarzschild, M. 1958, in *Stellar Populations*, ed. D. O'Connell, (Specola Vaticana), Ric A, 5, 59.

Searle, L. and Zinn, R. 1978, *ApJ* **225**, 357.

Searle, L., Wilkinson, A., and Bagnuolo, W.G. 1980, *ApJ* **239**, 803.

Secker, J. 1992, *AJ* **104**, 1472.

Shapley, H. 1914, *ApJ* **40**, 448.

Shapley, H. 1916, *ApJ* **43, 217.**

Shapley, H. 1918, *ApJ* **48, 81.**

Shapley, H. 1922, *Proc. Nat. Acad. Sci. Washington* **8**, 69.

Shapley, H. 1930, *Star Clusters*, (New York: McGraw-Hill).

Shapley, H. 1939, *Proc. Nat. Acad. Sci. Washington* **25**, 423.

Siegel, M.J. 1982, *PASP* **94**, 122.

Silbermann, N.A. and Smith, H.A. 1993, in prep.

Simon, N.R. 1988, in *Pulsations and Mass Loss in Stars*, ed. R. Stalio and L.A. Willson (Dordrecht: Kluwer), p. 27.

Simon, N.R. 1990, *ApJ* **360**, 119.

Simon, N.R. 1992, *ApJ* **387**, 162.

Simon, N.R. and Clement, C.M. 1993, ApJ 410, 526.

Simon, N.R. and Cox, A.N. 1991, *ApJ* **376**, 717.

Skillen, I., Fernley, J.A., Jameson, R.F., Lynas-Gray, A.E., and Longmore, A.J. 1989, *MNRAS* **241**, 281.

Smith, H.A. 1981, *PASP* **93**, 721.

Smith, H.A. 1984a, *ApJ* **281**, 148.

Smith, H.A. 1984b, *PASP* **96**, 505.

Smith, H.A. 1985a, *AJ* **90**, 1242.

Smith, H.A. 1985b, *PASP* **97**, 1053.

Smith, H.A. and Butler, D. 1978, *PASP* **90**, 671.

Smith, H.A. and Sandage, A. 1981, *AJ* **86**, 1870.

Smith, H.A. and Stryker, L.L. 1986, *AJ* **92, 328.**

Smith, H.A., Searle, L., and Manduca, A. 1988, unpublished.

Smith, H.A., Silbermann, N.A., Baird, S.R., and Graham, J.A. 1992, *AJ* **104**, 1430.

Smith, H.A., Matthews, J.M., Lee, K.M., Williams, J., Silbermann, N.A., and Bolte, M. 1994, *AJ* **107**, 679.

Sperauskas, J. 1987, *Vilniaus Astron. Obs. Biul.* Nr. 79, 36.

Sperra, W.E. 1910, *Astron. Nachr.* **184**, 241.

Stagg, C. and Wehlau, A. 1980, *AJ* **85**, 1182.

Stellingwerf, R.F. 1975, *ApJ* **195**, 441.

Stellingwerf, R.F. 1976, in *Multiple Periodic Variable Stars*, ed. W.S. Fitch (Dordrecht: Reidel), p. 153.

Stellingwerf, R.F. 1982, *ApJ* **262**, 330.

Stellingwerf, R.F. 1984, *ApJ* **277**, 322.

Stellingwerf, R.F. and Bono, G. 1993, in *New Perspectives on Stellar Pulsation and Pulsating Variable Stars*, ed. J. M. Nemec and J. M. Matthews, (Cambridge: Cambridge Univ. Press), p. 252.

Stellingwerf, R.F. and Bono, G. 1994, private communication.

Stellingwerf, R.F., Gautschy, A., and Dickens, R.J. 1987, *ApJL* **313**, L75.

Stetson, P.B., VandenBerg, D.A. and McClure, R.D. 1985, *PASP* **97**, 908.

Stobie, R.S. 1971, *ApJ* **168**, 381.

Stobie, R.S., Bishop, I.S., and King, D.L. 1986, *MNRAS* **222**, 473.

Storm, J., Carney, B.W., and Beck, J.A. 1991, *PASP* **103**, 1264.

Stothers, R. 1980, *PASP* **92**, 475.

Stothers, R. B. 1987, *ApJ* **319**, 260.

Strugnell, P., Reid, I.N., and Murray, C.A. 1986, *MNRAS* **220**, 413.

Struve, O. 1947, *PASP* **59**, 192.

Struve, O. 1950a, *Stellar Evolution: An Exploration from the Observatory*, (Princeton: Princeton Univ. Press).

Struve, O., 1950b, *PASP* **62**, 217.

Struve, O. 1951, *Proc. Second Berkeley Symp.* (Berkeley: Univ. California Press), p. 403.

Struve, O. and Blaauw, A. 1948, *ApJ* **108**, 60.

Stryker, L.L., Da Costa, G.S. and Mould, J.R. 1985, *ApJ* **298**, 544.

Stryker, L.L., Hesser, J.E., Hill, G., Garlick, G.S., and O'Keefe, L.M. 1985b, *PASP* **97**, 247.

Sturch, C. 1966, *ApJ* **143**, 774.

Sujarkova, O.G. and Shugarov, S.Yu. 1981, *PZ* **21**, 505.

Suntzeff, N.B., Friel, E., Klemola, A., Kraft, R.P., and Graham, J.A. 1986, *AJ* **91**, 275.

Suntzeff, N.B., Kinman, T.D., and Kraft, R.P. 1991, *ApJ* **367**, 528.

Sweigart, A.V. 1991, in *Precision Photometry: Astrophysics of the Galaxy*, ed. A.G.D. Philip, A.R. Upgren, and K.A. Janes, (Schenectady: L. Davis Press), p. 13.

Sweigart, A.V. and Gross, P.G. 1976, *ApJS* **32**, 367.

Sweigart, A.V. and Renzini, A. 1979, *A&A* **71**, 66.

Sweigart, A.V., Renzini, A., and Tornambe, A. 1987, *ApJ* **312**, 762.

Swope, H.H. 1967, *PASP* **79**, 439.

Szeidl, B. 1965, *Mitt. Sternw. Ungar. Akad. Wiss.*, No. 58.

Szeidl, B. 1975, in *Variable Stars and Stellar Evolution*, ed. V.E. Sherwood and L. Plaut, (Dordrecht: D. Reidel), p. 545.

Szeidl, B. 1976, in *Multiple Periodic Variable Stars*, ed. W.S. Fitch, (Dordrecht: Reidel), p. 133.

Szeidl, B. 1988, in *Multimode Stellar Pulsations*, ed. G. Kovacs, L. Szabados, and B. Szeidl (Budapest: Konkoly Observ.), p. 45.

Taam, R.E., Kraft, R.P., and Suntzeff, N. 1976, *ApJ* **207**, 201.

Taylor, P.O. 1977, *JAAVSO* **6**, 56.

Teays, T.J. 1993, in *New Perspectives on Stellar Pulsation and Pulsating Variable Stars*, ed. J. M. Nemec and J. M. Matthews, (Cambridge: Cambridge Univ. Press), p. 410.

Teays, T.J. and Simon, N.R. 1985, *ApJ* **290**, 683.

Thackeray, A.D. 1951, *Observatory* **71**, 219.

Thackeray, A.D. 1958, *MNRAS* **118**, 117.

Thackeray, A.D. 1974, *Mon. Not. Astron. Soc. South Africa* **33**, 66.

Thackeray, A.D. and Wesselink, A.J. 1953, *Nature* **171**, 693.

Thuan, T.X. and Gunn, J.E. 1976, *PASP* **88**, 543.

Tift, W.G. 1963, *MNRAS* **125**, 199.

Tsesevich, V.P. 1966 *The RR Lyrae Stars* (Kiev: Naukova Dumka).

Tsesevich, V.P. 1972, *Vistas in Astronomy*, **13**, 241.

Tsesevich, V.P. 1975, in *Pulsating Stars*, ed. B.V. Kukarkin (New York: John Wiley), p. 144.

Tsesevich, V.P. and Kazanasmas, M.S. 1963, *An Atlas of Finding Charts for RR Lyrae Stars* (Odessa).

Tuggle, R.S. and Iben, I. Jr. 1972, *ApJ* **178**, 455.

Tuggle, R.S. and Iben, I., Jr. 1973, *ApJ* **186**, 593.

van Agt, S.L. Th. 1967, *BAN* **19**, 275.

van Agt, S.L. Th. 1973, in *Variable Stars in Globular Clusters and Related Systems*, ed. J.D. Fernie (Dordrecht: D.Reidel), p. 35.

van Agt, S.L. Th. 1978, *Publ. David Dunlap Observ.* **3**, 205.

van Agt, S. and Oosterhoff, P. Th. 1959, *Ann. Sterren. Leiden* **21**, 253.

van Albada, T.S. and Baker, N. 1971, *ApJ* **169**, 311.

Van Albada, T.S. and Baker, N. 1973, *ApJ* **185**, 477.

Van den Bergh, S. 1957, *AJ* **62**, 334.

Van den Bergh, S. 1967, *PASP* **79**, 460.

Van den Bergh, S. 1993, *AJ* **105**, 971.

Van Herk, G. 1965, *BAN* **18**, 2.

VandenBerg, D.A. and Bell, R.A. 1985, *ApJS* **58**, 561.

de Vaucouleurs, G. 1993, *ApJ* **415**, 10.

Walker, A.R. 1985, *MNRAS* **212**, 343.

Walker, A.R. 1989a, *PASP* **101**, 570.

Walker, A.R. 1989b, *AJ* **98**, 2086.

Walker, A.R. 1990, *AJ* **100**, 1532.

Walker, A.R. 1991, in *The Magellanic Clouds*, ed. R. Haynes and D. Milne (Dordrecht: Kluwer), p. 307.

Walker, A.R. 1992a, *AJ* **103**, 1166.

Walker, A.R. 1992b, *AJ* **104**, 1395.

Walker, A.R. 1992c, *ApJL* **390**, L81.

Walker, A.R. and Mack, P. 1986, *MNRAS* **220**, 69.

Walker, A.R. and Mack, P. 1988a, *AJ* **96**, 872.

Walker, A.R. and Mack, P. 1988b, *AJ* **96**, 1362.

Walker, A.R. and Terndrup, D.M. 1991, *ApJ* **378**, 119.

Wallerstein, G. and Cox, A.N. 1984, *PASP* **96**, 677.

Walraven, Th, 1949, *BAN* **11**, 17.

Wan, L., Mao, Y.-Q., and Ji, D.-S. 1981, *Annals of Shanghai Observatory*, No. 2.

Wehlau, A. and Demers, S. 1977, *A&A* **57**, 251.

Wehlau, A. and Sawyer Hogg, H. 1978, *AJ* **83**, 946.

Wehlau, A., Conville, J., and Sawyer Hogg, H. 1975, *AJ* **80**, 1050.

Wehlau, A., Sawyer Hogg, H., Moorhead, R., and Rice, P. 1986, *AJ* **91**, 1340.

Wehlau, A., Nemec, J.M., Hanlan, P., and Rich, R.M. 1992, *AJ* **103**, 1583.

Weller, W., Mateo, M., and Krzeminski, W. 1991, *Bull. American Astron. Soc.* **23**, 877.

Wesselink, A.J. 1946, *BAN* **10**, 91.

Wesselink, A. J. 1969, *MNRAS* **144**, 297.

Wesselink, A.J. 1971, *MNRAS* **152**, 159.

Wesselink, T. 1987, Dissertation, Katholieke Universiteit te Nijmegen.

Westpfahl, D. 1992, preprint.

Wilson, L.A. and Bowen, G.H. 1984, *Nature* **312**, 429.

Wisniewski, W. 1971, in *New Directions and New Frontiers in Variable Star Research*, Veroff. der Remeis-Sternwarte Bamberg, IX, Nr. 100, 247.

Woolley, R. 1966, *Royal Observ. Annals* No. 2.

Woolley, R. 1978, *MNRAS* **184**, 311.

Woolley, R. and Davies, E. 1977, *MNRAS* **179**, 409.

Woolley, R. and Savage, A. 1971, *Roy. Obs. Bull.* No. 170.

Woolley, R., Harding, G.A., Casells, A.I., and Saunders, J. 1965, *Roy. Obs. Bull.* No. 97.

Yao, B. 1987, *ESO Messenger* **50**, 33.

Yi, S., Lee, Y.-W., and Demarque, P. 1993, *ApJL* **411**, L25.

Yoshii, Y. and Saio, H. 1979, *Publ. Astr. Soc. Japan* **31**, 339.

Zhao, J.-L. 1988, *Sci. Sin. Ser* A **31**, 734.

Zhevakin, S.A. 1953, *Russian A.J.* **30**, 161.

Zinn, R. 1980, *ApJ* **241**, 602.

Zinn, R. 1985a, *ApJ* **293**, 424.

Zinn, R. 1985b, *Mem. S. A. It.* **56**, 223.

Zinn, R. 1986, in *Stellar Populations*, ed. C.A. Norman, A. Renzini, and M. Tosi, (Cambridge: Cambridge Univ. Press), p. 73.

Zinn, R. 1993, in *The Globular Cluster–Galaxy Connection*, ed. G.H. Smith and J.P. Brodie, (San Francisco: ASP), p. 38.
Zinn, R. and King, C.P. 1982, *ApJ* **262**, 700.
Zinn, R. and West, M.J. 1984, *ApJS* **55**, 45.

Index

American Association of Variable Star Observers, 97
Andromeda Galaxy (M31), 8, 37, 118–19, 133–6, 138–9
antalgol stars, 4
Arequipa, Peru 2, 118

Baade–Wesselink method, 8, 18, 25, 28–9, 31–5, 37, 76, 87–8, 128
Baade's Window, 79, 81
Bailey types, 2–3, 15–16, 22–3
Beta Cephei variables, 5
Blazhko effect, 23–4, 83, 89, 91, 102–11, 117
  definition, 103
  explanations, 109–11
  frequency of occurrence, 104, 106–7
breathing pulses, 92

Canada–France–Hawaii Telescope, 136
Cepheids, 4, 9, 25–6, 117, 128
  anomalous, 119, 132–3, 137
  classical (Type I), 4–6, 25–6, 118–19, 132–5, 138
  period–luminosity relation, 4, 9, 26, 37, 118, 133–5
  type II (W Virginis), 2, 6, 26, 132–5
Cerro Tololo Interamerican Observatory, 121
charge-coupled devices (CCDs), 30, 39, 118–19, 121, 128, 136, 139
cluster-type variables, 2–4

ΔS method, 11, 16, 65–7, 73, 88, 140
δ Scuti stars, 5, 117
dwarf spheroidal galaxies, 8, 125, 129–34
  Carina, 130–2, 139
  Draco, 106, 109, 113, 115, 130–2, 134
  Fornax, 129–32, 134
  Leo I, 129, 131
  Leo II, 123–4, 129–31, 134
  Sculptor, 129–32, 134
  Sextans, 129–32
  Ursa Minor, 113, 130–2

[Fe/H], definition of, 10, 140
Fourier decomposition, 60–2

galactic center, distance to, 4, 77, 79
globular clusters, 22–3, 25–6, 34–7, 39–63, 119–21, 130–31, 134–6
  ages, 12, 26, 42–3, 62–3, 81–2, 120–1

astrometry of RR Lyraes in, 29, 37
color–magnitude diagrams, 13, 30, 41, 45–6, 60
discovery of variable stars in, 1–2, 39–40
dispersion in magnitude of RR Lyraes, 27–8, 48
and galactic evolution, 62–3
horizontal branch morphology, 28, 40–3, 59, 63, 71, 80
luminosity function, 31
numbers of RR Lyraes in, 2, 6, 39–40
R-method, 12
red giants, 31
(See also, instability strip, Magellanic Clouds, Oosterhoff groups, period changes, second parameter problem)
globular clusters; individual
  AM-1, 43
  Eridanus, 43
  Hodge 11 (LMC), 120
  IC4499, 113–15
  Lindsay 1 (SMC), 121
  M3 (NGC 5272), 2, 22, 34–5, 40, 42–6, 48–50, 53, 55–7, 63, 99, 101, 104–5, 113–14, 116, 122, 124, 126, 129–30, 138–9
  M4 (NGC 6121), 48, 99
  M5 (NGC 5904), 2, 30, 49, 53, 87, 99, 105–6, 113
  M13 (NGC 6205) 59
  M14 (NGC 6402), 99
  M15 (NGC 7078), 23, 34–5, 44–6, 48–50, 55–7, 59–60, 62, 99–101, 106, 109, 111–16, 126, 129–30
  M22 (NGC 6656), 99, 101
  M53 (NGC 5024), 49, 99
  M62 (NGC 6266), 53
  M68 (NGC 4590), 62
  M80 (NGC 6093), 1
  M92 (NGC 6341), 87, 99
  M107 (NGC 6171), 99, 136
  NGC 121 (SMC), 119–22, 124
  NGC 288, 43
  NGC 339 (SMC), 120
  NGC 362, 43
  NGC 1783 (LMC), 122
  NGC 1786 (LMC), 120
  NGC 1835 (LMC), 120
  NGC 1841 (LMC), 120
  NGC 2019 (LMC), 120
  NGC 2210 (LMC), 120, 122
  NGC 2257 (LMC), 99, 119–22
  NGC 3201, 45

Printed in the United States
By Bookmasters